**Unique Methods for Analyzing Failures
and Catastrophic Events**

Unique Methods for Analyzing Failures and Catastrophic Events

A Practical Guide for Engineers

Anthony Sofronas

Registered Office
John Wiley & Sons, Inc., 111 River Street, Hoboken, NJ 07030, USA

Editorial Office
111 River Street, Hoboken, NJ 07030, USA

For details of our global editorial offices, customer services, and more information about Wiley products visit us at www.wiley.com.

Library of Congress Cataloging-in-Publication Data

Names: Sofronas, Anthony, author.
Title: Unique engineering methods for analyzing failures and catastrophic events: a practical guide for engineers / Anthony Sofronas.
Description: Hoboken, NJ : Wiley, 2022. | Includes bibliographical references and index.
Identifiers: LCCN 2021055315 (print) | LCCN 2021055316 (ebook) | ISBN 9781119748250 (cloth) | ISBN 9781119748267 (adobe pdf) | ISBN 9781119748274 (epub)
Subjects: LCSH: Failure analysis (Engineering) | Production engineering.
Classification: LCC TA169.5 .S64 2022 (print) | LCC TA169.5 (ebook) | DDC 620/.00452–dc23/eng/20211123
LC record available at https://lccn.loc.gov/2021055315
LC ebook record available at https://lccn.loc.gov/2021055316

Cover image: Wiley
Cover design: © Dimitris66/Getty Images, Image courtesy of Anthony Sofronas

Set in 9.5/12.5pt STIXTwoText by Straive, Chennai, India
SKY10034348_050522

To my Lord who made this all possible and to my wife Cruz and my family

Contents

About the Author

Anthony Sofronas is an engineer, author, educator, and consultant. He has extensive practical experience in designing and troubleshooting machinery, structures, and pressure vessels. His 50 years in the industry have been with the General Electric Company, the Bendix Corporation and his most recent 25 years in industry with the Exxon Mobil Corporation. In this position, he had worldwide responsibilities with his team troubleshooting equipment failures. Since retirement, he has embarked on a career as an author, lecturer, and engineering consultant. He has authored over 120 technical articles and has written four engineering books based on his work. He has presented seminars in eight countries and consulted worldwide.

Dr. Sofronas graduated from the University of Detroit with a Doctor of Engineering, Pennsylvania State University, Masters of Engineering, Northrop Institute of Technology, Bachelors of Science in Mechanical Engineering and New York State University at Farmingdale with an Associate of Applied Science in Mechanical Power Technology.

He has been registered as a Professional Engineer in Texas and elected to the honor society Tau Beta Pi. He received the Society of Manufacturing Young Engineering award in 1980 for his research work, which was considered a pioneering analytical work, into the drilling process.

Preface

Technical specialists and investigators in all fields review case studies to observe ways to solve problems, especially failures. The immediate problem that needs to be solved may be far from that of the case study but the example provided might suggest a unique way to analyze a failure. A comprehensive solution may be more accurate, but reducing the complexity can make the failure cause understandable, faster to implement a repair, and easier to explain to management.

During my mechanical engineering career, I've worked on all types of failures, learning, and documenting something important from each one. Much of this knowledge was based on being curious. Many problems were solved without compensation. This is because someone asked a question and performing an analysis would provide an answer. Many of these types of solutions were used in some other form to analyze a failure and are presented in this book.

Most of these solutions were used in a working plant environment where there was always pressure for an immediately solution. Downtime was directly related to lost production and lost profits. Ensuring a safe start-up and operation was always a major consideration. Most examples include the basic calculation methods and equations needed to solve the problem. A summary at the end of each section highlights the key points presented and lessons learned. Chapter 12 is a discussion on some of the equations used.

Some of the examples show the use of certain techniques. For example, they show how to calculate impact forces by describing the force of a bat hitting a baseball. This was later used to explain the force of a chunk of a product hitting an impeller and breaking bolts. Since there may be no data to support the analysis method, the ball/bat analogy and analysis can confirm the impeller analysis is reasonable.

The intended audience for this book is practicing engineers, technicians, and scientists desiring to see how engineering tools can be used to answer questions. It will also be useful to anyone who wishes to see how practical problems are solved, such as managers, supervisors, attorneys, or students who are involved with technical issues.

The cover design represents the flow dynamics used in a finite element analysis used by the author along with some of the equations used in this book.

Acknowledgments

First I wish to thank my dear wife Mrs. Cruz Velasquez Sofronas for putting up with all my technical terms discussed with her over the years. She has actually started to understand them. She has been so patient and helpful in suggesting better wording for many of the sections.

This book has been a real family affair. Our son Steve and Yuki and our grandchildren Ally and Anthony helped with the artwork and models, and our daughter Maria and David helped by answering my many graphics questions. I thank them all.

I wish to thank Mr. Heinz Bloch, a prolific writer, educator, and friend for suggesting that I write this book along with his many beneficial suggestions.

I thank Mr. Richard S. Gill, my colleague and friend for his encouragement and who I worked with on many of these cases. Many enjoyable hours of technical discussion with him helped in solving the actual problems.

Many thanks go to my colleague and friend, Mr. Geoffrey Kinison who saw the worth of this book and provided many useful contributions.

My appreciation also goes to the many technicians, operators, machinists, engineers, and others who have also contributed to this work in some way.

My thanks also go to John Wiley & Sons, Inc. especially Ms. Summers Scholl for seeing the value of this work and agreeing to follow it through the publishing process.

1

Engineering Suggestions Based on Experience

A Failure Cause That Hasn't Been Identified Can't Be Rectified

1.1 What Should We Learn from This Book?

This is a very important question we should ask ourselves before using our valuable time. After all, there are many other sources of information available to us. So what makes this book so special? Simply that many of these solutions are those of the author and have never been available to the public before. They have been used to solve actual problems in industry.

Analyzing failures is a gratifying part of a technical person's career, some engineers will say the most interesting and rewarding. This is because it can take you into many areas, all quite different. Various areas of your technical background can be used in solving a problem. During my mechanical engineering career I've worked on all types of failures, learning and documenting something important from each one. Much of my knowledge was based on being curious.

Actual examples and experiences are always valuable, especially when they can show how to simplify and solve difficult problems. While there are many sources of information, it is the personal experiences that make this book unique.

One wonderful thing about analyzing something with mathematics is that once you have a viable model you have developed a time machine of sorts. You can analyze what a piece of equipment has done in the past and what it will do in the future. For example, with fracture mechanics you can determine when a crack formed and how long it will take to propagate to a failed condition. A small crack in a gear tooth and a small crack in a coupling are actual examples. You don't have to know the initial crack size, you can just say it's 1/64 in. long and see what the cyclic stress does to elongate it to a failure length.

Unique Methods for Analyzing Failures and Catastrophic Events: A Practical Guide for Engineers, First Edition. Anthony Sofronas.

In this book, many different areas are investigated. The purpose is to show how one area in engineering can affect another. Heat transfer, dynamics, and kinematics are used to describe the human body as in Chapter 8. Strength of materials, dynamics, and fluid flow are used to explain various Earth-related phenomena as in Chapter 9.

Copying what nature or other creatures do and trying to mimic them, such as flying, is called the science of biomimetics. The first example that comes to mind is Leonardo da Vinci. He watched and sketched the flight of birds. He never built a flying machine but he realized the science involved. Section 4.5 shows the analytical flight of aircraft and bumblebees. "Walking on Water" is another observation made of a Basilisk lizard and researchers have tried to duplicate this technique. It is shown in Section 10.13 and explains why this is so difficult for an unpowered human to achieve.

The analysis methods used are not meant to be highly accurate, but accurate enough, believable, and understandable enough to solve a problem quickly.

The versatility one can have in the field of science and engineering is immense. Most practicing engineers and scientists know this but for those trying to decide on an area to go into, the following discussion may be useful.

I've been a practicing mechanical engineer for over 50 years and during this time, in addition to getting married, sending my wife and children through college and traveling with them around the world, I've been:

- Manager of advanced engineering
- Worldwide problem solver
- Contributing editor to a technical magazine with 110 articles
- Researcher who developed a new drill and sampling device for a Martian lander
- Wrote four technical books
- Worked on machinery, pressure vessel, and structural problems of all types
- Taught in a university
- Owned a consulting firm after retirement
- Presented technical seminars to industry worldwide
- Manufactured and sold a motorized tow and vibration indicator for aircraft

My love of mechanical things also allowed me to become a pilot of my own aircraft so I could fly to consulting jobs if I so desired. I also had time to restore my favorite things such as old cars and Jeeps.

I graduated from a 2-year technical college because I loved working on machinery. The design part was extremely interesting to me so I went on for my mechanical engineering degree. I'm sure I could have had a very successful career with this 2-year degree; however, the engineering degree offered many additional opportunities. My only concern was that hands-on engineering might not be possible and I'd be confined to working from behind my desk. This never occurred and 50% of the time was out of the office. Usually, the out of the office was spent at other sites both worldwide and in the United States analyzing failures or reviewing designs with manufacturers. Attending seminars, consulting with field personnel, and visiting outside machine shops and metallurgy laboratories are some of the other activities. There were many trips out of the country and I was able to work on many of the continents. Obtaining an engineering degree requires hard work, but then anything worthwhile often does.

The analytical solutions used in this book are not meant to be highly accurate as many assumptions are made. They are intended to provide data to solve actual problems when little data is available. In this way, they have helped a busy engineer make critical decisions on in-plant problems with minimal risk. It is hoped it will allow the reader to do the same.

1.1.1 Summary

Engineering is extremely versatile and flexible. Advanced degrees can greatly widen one's career opportunities [1]. This book should help the readers by allowing them to mathematically analyze the causes of a failure, and allowing them to provide the most likely cause. This will greatly reduce the personal risk when they are required to provide a solution.

Reference

1 Sofronas, A. (2016). *Survival Techniques for the Practicing Engineer*. Wiley.

1.2 We All Contribute to Each Other's Success

At the start of my career with the help of an excellent mentor, I became quite good at analyzing equipment and structures using analytical techniques. There was a point where I thought that anyone in mechanical engineering who didn't perform detailed calculations wasn't doing real engineering work.

Of course, this was my youthful arrogant thinking and when I realized how absurd it was I repented and was ashamed. I noticed that those I worked with had unique talents that I didn't have. Not having recognized that and my career would have been much less productive. As I have told people "Some of my best works were achieved with the efforts of others."

A way to help explain this is with a few examples.

Example 1.1 A newly installed vibrating conveyor was experiencing extensive cracking and had been noticed by the unit's machinist. Arlon shared with me that he thought the spring hangers were misaligned and measurements confirmed this. We used the measurements to build an analytical model that predicted high stresses and cracking of welds due to the misalignment [1]. Only after reviewing the measurements and calculations did the manufacturer agree to perform the repairs and tighten the tolerances at their cost. Recognizing his knowledge of the equipment, the machinist was asked to supervise the job which benefited his career. All would be much more difficult without his expertise.

Example 1.2 A large gas-engine compressor experienced a crankcase explosion at a remote compressor house site in Louisiana. I was called to determine the cause. A technician who had 20-year experience repairing these large compressors was brought along to help. Crawling around inside the engine he showed me tricks of the trade. He tapped critical bolts with a hammer to see if they were loose. In this way, he located several loose and broken bolts. It seems a cylinder was loose and it rubbed the piston skirt which caused a hot spot to develop. This ignited the oil vapors in the crankcase. Don's practical knowledge of the compressor, being more extensive than mine, had located the cause.

Example 1.3 I worked for a company that was repowering ships with a 3,000-horsepower diesel engine they manufactured. My assignment was to travel the Mississippi river on this seagoing tug to monitor the operation since I would be performing the torsional analysis of the new system. Management wanted to know if this was a good application for our engine. Half-way through the journey, one of the propellers hit a log and broke off a couple of blades. The captain continued on to deliver the barges, which took several hours. During this time, the ship was shaking excessively from the unbalance forces. The rubber-type coupling connecting the engine to the gearbox was smoking so badly that the crew put water on it to cool it down. An experienced crewman said this had happened before. His experiences allowed us to insert a clause in the warranty that protected my company should a similar condition occur. Sometimes witnessing a failure is a good thing.

Example 1.4 A friend was designing a gearbox to adapt to a high-performance automotive engine for use on experimental aircraft. I asked him if he had performed a gear train analysis on the gearbox and bearings. Bud said that the gears and bearings had been used at lower speeds and at a horsepower twice what he would be using. This fellow was a master machinist and mechanic who had been involved in all sorts of racing activities during his lifetime. Higher horsepower at lower speeds meant higher gear loading and for his gearbox provided an adequate design margin. He didn't realize this but his wealth of experience told him it would be acceptable. A subsequent analysis showed he was correct in his thinking. Here is a non-engineer whose input I would value because of his breadth of experience based on trial and error and his logical thought process.

These examples show how input from experienced resources is so important. Of course, for your own protection, you need to screen the data yourself and don't blindly accept input as being true.

1.2.1 Summary

Utilizing others who are experts in their respected areas will allow you to better utilize your own skills. Be sure to give them credit for their contribution.

Reference

1 Sofronas, A. (2006). *Analytical Troubleshooting of Process Machinery and Pressure Vessels*, 317. Wiley.

1.3 Why Performing Calculations is Important to an Engineer's Career

This is about the value engineering calculations can have in helping an engineer's career by controlling the risks taken.

Most engineers realize they can proceed through their careers doing very few engineering calculations. Managing projects or people, selecting and procuring equipment, working with manufacturers and consultants to design products, troubleshooting equipment, pressure vessels, or structures, or using sophisticated computer programs may not require engineering calculations. This type of work can be done by talented personnel capable of making sound decisions based on the data they obtain.

However, by using engineering mathematical techniques, you can greatly enhance your abilities since you can generate the data needed. You don't have to rely solely on someone else's data which might be flawed.

Early in my career, I realized that my managers didn't want to know the details on how I arrived at my results, which were usually from an analysis I was quite proud of. What they noticed was that I was willing to make difficult decisions, meaning take a risk, solve the problem expediently, safely, and cost effectively. Usually, it meant the system design or equipment was now operating successfully.

Lack of their interest in the calculations didn't bother me since the analysis allowed the following:

- To perform a logical analysis and provide documentation that supported my decision and made me feel comfortable with the risk taken in the decision.
- To use historical failure data to analyze new designs and prevent the same type of failure from occurring again. When smaller units aren't failing but larger ones are an analysis can show why.
- To verify the analysis by using other actual failures to see if the results are the same. Cracks and failure zones show you where to center your analysis.
- To feel confident reviewing any type of unique equipment because it could be simplified, analyzed, and understood. The internal stresses, dynamics, and other phenomena are understood, something that others may not have known about. Using equations, the system can be modeled and what is happening inside an operating system observed.
- To calculate if and when something is going to fail. Analysis is exciting and like time travel. It allows you to look at a failure before it occurs or determine when it is going to fail in the future.

This will allow you to develop a strong track record of successful risk taking and decision-making and become known as someone who can help resolve difficult problems. This is what will help you advance and be recognized by management not by them having to see and understand your calculations.

Some simple examples will help explain this. Many more are in the references.

Example 1.5   Why Did the Shaft Twist?

A 20-ft long extruder shaft made of a 316 stainless steel had twisted along its length. In a meeting that was discussing the failure, there was much speculation as to what or who was the cause. It was asked if this was the original shaft. The answer was yes and with further questioning learned the motor had been increased from 800 to 1,200 horsepower several years earlier to handle a tougher product (see Table 1.3.1).

The following was put on a flip chart in the meeting [1]:

Yield strength (σ_{yield}) for 316 steel = 35,000 lb/in.^{2}

Shear stress in shaft due to operating loads (σ_{shear})

$$= 321{,}000^*HP/(D^3{}^*RPM)$$

Table 1.3.1   Data put on flip chart.

Horsepower $D = 4$ in., RPM = 200	σ_{shear}	$\sigma_{yield}/\sigma_{shear}$
800	20,000	1.75
1,200	30,000	1.17

Notice that the shaft would have required almost double the load to fail at 800 horsepower but much less at 1,200. This is why it never failed at 800 horsepower. There was enough margin to handle periodic adverse operating conditions, such as cold product start-ups. A detailed analysis was performed and the shaft was replaced with a higher yield strength 17-4 material.

This quick calculation changed the direction and results of the meeting. A repeat failure would likely have occurred with continued operation with the spare 316 shaft.

Example 1.6   Can We Start The Unit Up?
A wrong answer based on someone else's opinion or erroneous data with a subsequent failure can jeopardize your career. This is where your ability to do calculations can be important. In this case, a large heated rotating disk dryer operating at a plant [1] had developed cracks at welds. Several similar smaller units hadn't developed cracks. The problem was compounded since a new much larger scaled-up unit had been shipped overseas and was waiting for installation in a new facility. The question from management was "What Should We Do Next?"

A simple finite element model was used to determine the stresses at the weld along with a cyclic crack growth analysis on the cracked dryer. After having performed these calculations on all the dryers the engineer felt confident in saying, "Based on the analysis if the new unit is put into service it will crack and fail in less than one year." It was recommended that the unit be sent back to the United States from Asia and strengthened by replacing only the first 3 disks out of 40 with thicker ones to expedite the repair. The analysis indicated that this was all that was required, along with better welding methods. Project management and the dryer manufacturer who was reluctant to modify the new unit agreed after reviewing the calculations. The decision was a low-risk one for the engineer since the same analysis was used on the smaller units and it was now known why they hadn't cracked. When you have good data from calculations, your position will usually be the strongest. Others probably have no data to contradict yours and aren't willing to take the risk. Here is a case where an analysis predicted the future and a crack growth calculation was a powerful tool.

These types of calculations are well within the capability of most engineers who look for the opportunity to utilize them. We need to recognize that this will eventually benefit our career by allowing us to take measured risks that others aren't willing to take. We could do this because we had data that others didn't. Calculations usually win over speculation when making recommendations.

1.3.1   Summary

By performing basic calculations an engineer will have new data that wasn't available. This will allow informed decisions to be made on equipment that is a first-of-its-kind design or one that has had repeated failures. A wealth of historic experience isn't necessary as new data can be calculated.

Reference

1 Sofronas, A. (2006). *Analytical Troubleshooting of Process Machinery and Pressure Vessels*, 33–315. Wiley.

1.4 How an Engineering Consultant Can Help Your Company

While engineering consultants are available for a wide range of client requirements such as developing or reviewing specifications for new equipment, this section is only concerned with analyzing equipment failures.

When a company has limited engineering resources, the failure of a critical piece of equipment and a quick resolution as to the cause can be demanding on the staff.

A consultant is a partner of the owner's failure analysis team providing valuable data to help in identifying, explaining, and addressing the failure causes.

Unique equipment failures can be difficult to troubleshoot since they may never have been experienced by the staff before. For example, a wrecked compressor housing as shown in Figure 1.4.1 or cracked welds on vibrating conveyor might be considered unique while a bearing or seal failure may not be.

One method to address the failure cause is to contact the equipment manufacturer to see if they can help provide a solution. There may be a reluctance on their

Figure 1.4.1 Wreck of a gas-engine compressor.

part if it could involve a costly warranty repair or could result in legal actions to them because of production losses or safety issues.

Often heard responses from the manufacturer are that:

- the correct operating procedures haven't been followed.
- the machine has been overloaded or abused.
- preventative maintenance hasn't been carried out.

These responses aren't helpful in getting the equipment back into operation and the owner may not agree with the manufacturer's statements.

The owner may ask for the design calculations in the failure area. The manufacturer may reply that it's proprietary information.

On new designs of larger power and geometric size, the manufacture's smaller and highly reliable unit may have been scaled up. Failures can occur in areas that don't scale up well such as welds and weld defects.

The following are some failures the author has worked where there was little design information available from the manufacture or the equipment was the first of its kind to the client company:

- vibrating conveyor structural failures
- material recycling choppers and grinder structural failures
- rotating mixer/dryer structural failures
- gearbox movement, gear pitting, and breakage
- extruder screw wear and breakage
- screw conveyor auger shaft failures
- agitator shaft, seal, and blade failures
- wrecks of gas-engine compressors
- screw compressor thermal rotor rubs
- cracked piping due to unknown vibration sources
- system failures due to torsional vibration

The consultant may have worked on similar failures for other companies and therefore has a wider experience base.

A consultant skilled in simplifying complex machines may develop an analytical model to better understand the possible causes. This allows the loads affecting the failure zone to be analyzed and explained in detail.

One output from such models is to determine the expected life in years due to the loads, stresses, and production rate. Defining methods to reduce these loads and stresses can increase the equipment's life.

Providing this type of data to the manufacturer will let them know you have detailed knowledge of the failure. With this they may be willing to share information on what can be done.

Experienced consultants are not inexpensive. The consultant must maintain the software and experience that a company engineer may not have. Finite element analysis (FEA), crack growth calculations, structural analysis, vibration analysis programs, and testing equipment are a few items necessary to do their job. Insurance, travel expenses are other necessities that must be included in the hourly billing.

The consultants billed work should be compared with:

- the production losses, legal liabilities, and repair costs of a repeat failure.
- the use of your engineering staff's time needed to do such an analysis.
- sharing the billed work with your company's other sites that may have experienced similar failures.

An engineering consult can be considered as an experienced temporary resource that is available at any time to help in your company's failure investigations.

1.4.1 Summary

Engineering consultants may have a wider base of experience from working with various industries. A similar failure may have been experienced at some other company. A similar approach can be used at your company.

1.5 · The Benefit of Keeping Complex Problems Simple

When you are writing a technical paper for the government, a doctoral dissertation, or a consulting job, keeping things simple certainly isn't recommended. These types of jobs require a detailed analysis and they are expected from you. When you have a $100,000 budget to perform an analysis, a high-level technical analysis method is expected. It will be written for technical people who understand the complex mathematics and computer software used in the report. Finite element and structural design programs are examples. They will handle difficult structural design, heat transfer, fluid flow, and vibration problems that are difficult to analyze.

For most of my career, simplifying complex problems was important. Much of this has been in a plant environment and then as a consultant to industry analyzing failures. Something has failed and you are called in to determine why and to help fix it so it won't happen again. When the downtime is costing the company thousands of dollars a day, there's little time for long-term procrastination on an analysis. Sometimes all the analysis has to do is satisfy the investigating team with the probable cause or causes and how to address them so they don't happen again. As you will see in Section 7.1 of this book, this can become a little involved but can be done relatively quickly.

Any analysis is only worthwhile when the following are fulfilled:

- The outcome agrees with the failure data. For example, if the failure of a shaft is a fatigue failure and a materials expert says it was an instantaneous failure, someone is in error. The analysis should agree or confirm the observed data.
- It can be used to implement a fix. This means there must be results that can be used. When a shaft fails in torsion, the analysis should be able to illustrate a possible cause such as the start-up of an extruder with cold product in it. In that case, "bump-starting" to free the plug should be cautioned in the start-up procedure.
- The results of the analysis should be understandable. Complex analyses usually have nondimensional graphs and many technical terms and equations that can't be understood by the practicing technical people, nor do they desire to. A clear deliverable with pictures of what has occurred and what needs to be done is what is needed.

Very rarely was the investigation team or management interested in how the failure problem was solved, meaning analyzed. All they want to know is, "What do you think caused it and what should we do?" This requires a very straightforward answer.

The finite element and structural programs can be used in simplified ways. For example, making a three-dimensional problem into a two-dimensional 1/4 axis-symmetric model in stress analysis, heat transfer, and fluid flow can greatly reduce the complexity, expense, and time. This makes them applicable to plant problems requiring immediate answers.

Here are some examples on where this was used.

Example 1.7 Reactor with Restrictor Head

This analysis was done to reverse a decision on installing a restrictor plate in a reactor. The plate was supposed to increase the flow velocity to reduce fouling.

This is a computational fluid dynamics (CFD) analysis of a reactor with a restrictor head plate. This 2D simplified analysis shows the high velocity in the 1-in. gap (c) region. It also shows the poor distribution and low flow in the inner tube.

Notice the simplification here. If you sweep the tubes, you have channels not individual tubes, so the results are only approximate, but did indicate the problem. Sometimes the engineer has to be creative in order to make timely decisions (see Figure 1.5.1).

Example 1.8 Extruder Barrel Distortion

A 50-ton extruder gearbox was moving on its base, even after careful alignment and bolt tightening. The simple "block" thermal stress FEA was performed.

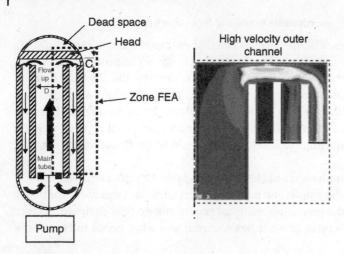

Figure 1.5.1 Reactor restrictor head.

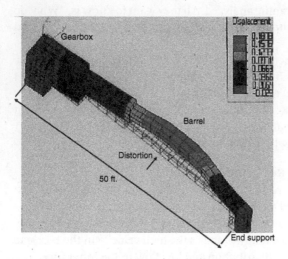

Figure 1.5.2 Extruder barrel distortion.

It showed that the fouled cooling passages in the barrel caused barrel distortion. This resulted in a thermal moment that would move the gearbox. The analysis was verified with laser measurements and thermal imaging. A better water passage cleaning schedule and temperature instrumentation solved the problem (see Figure 1.5.2).

Example 1.9   Aircraft Wind Screen Bird Strike
This was a study that reviewed the use of a thicker one-piece polymer wind-screen on an aircraft. Bird strikes can cause damage such as shown, so the

FEA stress analysis looked at the stresses produced using different thickness windscreens.

The analysis required the impact loads to be calculated analytically and applied to the 3D FEA model. With large bird impacts, it was impossible to keep the windscreen from failing no matter what reasonable thickness was used. The solution is to stay away from birds (see Figure 1.5.3).

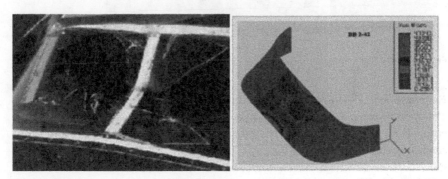

Figure 1.5.3 Aircraft windscreen bird strike.

Example 1.10 Cracking of a Cold Cycling Valve

This is a cycling valve spool piece that developed cracking over several years. An axis-symmetrical thermal stress analysis verified that it was due to the thermal cycling stress. Notice how simple the model is and that the steam jacket didn't have to be included, only the boundary temperature of the steam. Notice how cracking only developed in the high-temperature zone where the temperature cycled from −40 to +250°F (see Figure 1.5.4).

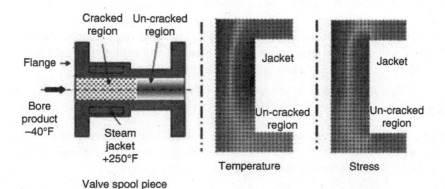

Figure 1.5.4 Cold cycling valve.

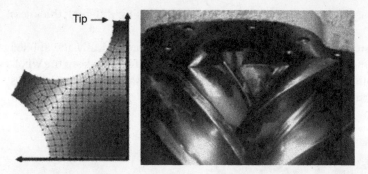

Tip →

Figure 1.5.5   Rotor tip rub.

Example 1.11   Tip Rub of Rotary Screw Compressor
This represents a 1/4 axis-symmetric coupled thermal stress analysis of a screw compressor rotor. The rotor was overheated for a short period of time and it was requested that an FEA be done to see if rotor tip contact was possible due to rotor growth. It was (see Figure 1.5.5).

These simplified models with results took less than one day each. The color version of the results was clear and easy to explain to the responsible parties and the solutions were made obvious. These types of methods were very approximate and certainly wouldn't stand the test of a technical evaluation with all of the approximations made. I would not want to defend them in a doctoral dissertation review. They did successfully solve the problem and have the plant back in service so everyone could get on to their next assignment.

That's what most of this book is about. Most engineers would not care about a bird impacting an aircraft windscreen. They might be interested in using the finite element method (FEM) on a similar failure, possibly the windows of a high-rise top floor window. The impact might not be from a bird but from wind debris. The restraint conditions would show if the window would "pop-out" or "pop-in."

Even if someone is not knowledgeable in FEA, knowing that the model can be simplified will allow the technical person to mention and explore this approach with a technical consultant. The saving in time can be substantial.

1.5.1   Summary

An easy-to-explain analysis will usually be more expedient to implement a solution than a complex one. It will allow the responsible parties to understand the cause better and implement solutions with confidence.

1.6 Taking Risks and Making High-Level Presentations

As engineers we like to limit our risks. As a general aviation pilot and a mechanical engineer this has served me well over the years. I didn't do things that were too risky and always had a couple of alternate plans in case something went wrong. My requests for a design modification were always supported with adequate calculations. Usually, when someone has performed a reasonable analysis their arguments carry more weight than those who are speculating on the cause, with no supporting data.

There can be problems with this approach.

The first is that there is always some risk involved in every engineering decision and you cannot progress far in your career if you are unwilling to take calculated risks. Consider a large steam turbine vibrating slightly above normal levels with blade fouling thought to be the problem. Management wants to know if they can run one week until a planned outage with that level of vibration. Your career will not be enhanced if you say it has to be shut down immediately, with no supporting data. Likewise this is not the time to try your first attempt of online washing of a steam turbine while it is in operation. This is a risky business if you have no experience or operating guidelines to follow for this procedure. This would be a good time to monitor the vibration level, talk with the manufacturer and others with similar machines and determine the risk in just monitoring the vibration levels. Defining at what level it will have to be shut down will still require some risk, but now others are involved. Obviously, there is much more but this illustrates the need for some risk. You can expect to make some judgment errors but they should not occur early in your career or consecutively.

Correctly communicating to management what has occurred and what needs to be done is so important to an engineer. It would be wonderful if engineers had the verbal ability of attorneys in presenting data to management. An attorney job is to make juries feel comfortable with what they are telling them and the decision that has to be made with the evidence and data they have. Unfortunately, many of us don't have this type of training. Fortunately, it can be learned by experience and watching other successful engineers. Your company's senior technical personnel didn't get where they are via a lack of communication skills or poor judgment.

There are three things I have found important when discussing work with senior management. The first two items are self-explanatory but the last will require an example.

1. Management does not like to hear bad news, so present positive plans.
2. Management does not want to hear a wish list of solutions so present only your best and the most cost-effective choice.

3. Management is not impressed with complex analysis or technical terms so the engineer must simplify the cause, solution, and implementation so it can be understood and acted on.

The management team you are presenting to may not be familiar with mechanical engineering and may have their expertise elsewhere. It's useful to adjust what you are presenting to suit your audience.

Suppose you are discussing the resonance of a structure and its failure. Now as engineers we know resonance can be a highly damaging vibration caused by exciting a structure's natural frequency. Resonance could be clearly demonstrated by bringing a tuning fork to the meeting, striking it and showing the resonance of the tuning fork. With no continued exciting input, meaning striking it, the vibration dies out. However, with a continued input and without material damping the tuning fork would fail in fatigue. The fatigue failure part could be demonstrated by bending a paper clip back and forth to demonstrate and explain fatigue. At any point, you can stop bending but some of its life has been used up. This is shown in Section 8.2. Before using these techniques, it is important to know how the management team will perceive this type of explanation. Is it too simple or is it helpful?

When preparing for a technical presentation, one approach used is to make a list of all the questions you may be asked during the meeting and research them thoroughly. This allows the best possible answers for the audience you are presenting to.

1.6.1 Summary

Taking risks in engineering is necessary in solving problems. These risks can be limited in their severity by having good data either obtained or developed with analysis.

1.7 Searching the Literature for Data

Gathering accurate data is essential for failure analysis and other types of research. Writing this book required considerable research and some of the methods used are outlined here.

It's relatively easy to do a literature search today. Before electronic methods were available, informed individuals, the library, periodicals, and technical journals usually provided the results needed. Science, engineering, and technology centers provided research in most areas.

In addition to the above sources, much of the data utilized now is located on the Internet. The data can be used to verify analytical models, develop new ones, or read about lessons learned.

Data is also available as unclassified government reports on the Internet. When you use the correct search topics, they will usually show up. For example, Ref. [1]

was found searching under NASA REPORTS. Such reports also contain valuable references and test data to help you widen your search. The results are quite reliable since these types of reports have been thoroughly reviewed. When a document has multiple contributors, the review process is even more thorough. Unfortunately, brevity and simplicity do not usually prevail in such documents.

There can be problems with some of the sites on the Internet and other social media sources providing information. Many have been found to contain erroneous information which could be dangerous if believed and used.

When a site has an experienced contributor who can comment on erroneous data the site holds more value. When someone is using questionable language, give the site a low score on believability. The site probably is not being screened for substance. Here are some other problems I've noted on some Internet sites.

1.7.1 Equations

Even in my own works, when an equation is converted from one type format to another errors can occur. For example, when converting from a Word document to a PDF document, some of the symbols in the equations may not be recognized. Always verify the equations you use and ask for a final review when they are converted. It's also useful to do a dimensional check on the equations to ensure the units are correct.

1.7.2 Facts

One person's statement of a fact may only be someone else's opinion and that opinion may not even be their own. What they are citing may have been obtained from another source and that source may have been incorrect. When researching for an article on pneumatic testing I've read, "If it blows up just get out of the way". My first thought was that this would be hard to do when fragments are flying at 2,500 ft/s. People want to have something to say but without good data it's just an opinion.

1.7.3 Credibility

You may know nothing about the person who is presenting the information. At least in a technical publication, the background of the author is known. A contributor may have valuable information based on their experiences, however, you won't be able to verify its value unless you cross-check it against other sources.

1.7.4 Accuracy of the Data

Much of the work I do doesn't have to be highly accurate. Getting the unit back-up and operating safely and cost-effectively along with eliminating the possibility of a repeat failure was the goal. Usually, the details on how this was done are not as important to management as that it got done safely and cost-effectively.

An overload can be 300% or 800% and break gear teeth. The exact percentage isn't as important as knowing why there was an overload and what should be done to keep it from reoccurring.

This is the reason obtaining test data from the literature is so important. Here are some additional suggestions for obtaining information:

1.7.5 Sources to Search

For locating test data for comparison with calculated data, one source is technical journals of societies that specialize in that type subject matter. Various journals such as published by the American Society for Mechanical Engineers usually have papers on most subjects in mechanical engineering. Other technical societies do the same. Government research publications usually contain test data. I have also used online Masters and Doctoral thesis of specific topics that will show up if you are persistent enough.

For general subject matter or information on scientists and people of interest, I have always found Wikipedia on the Internet and other such sources to be quite useful. When the information is incomplete, they will mention this. Usually, these have a good number of references to follow up on. For example, I searched Amazon books for [2] and then got it as an interlibrary loan.

It's amazing how much information there is online for learning different subjects. I taught myself CFD for use with the ANSYS program I had with online courses by just searching the web. Khan Academy [3] is an online source and a wonderful learning aid for brushing up on statistics and other subjects.

Your own experiences are a valuable source of information. I've made it a habit to write down sources used and where to get this type of data again. It may have been at a trade show, a lecture, reading a technical journal, book, or article or asking someone a question in a machine shop. Anything of importance would be saved. Most of my books have been written using this type of material. One good technical paper can generate many valuable references.

1.7.6 Summary

Advice to those new to engineering or the sciences is to record all the work you do. Especially important are failures that you have witnessed because eventually it will be useful to you. Most companies don't like to admit their failures and how they solved them, so this type of material is scarce.

References

1 Columbia Accident Investigation Board (2003). *National Aeronautics and Space Administration*, vol. 1, 248. Government Printing Office.
2 Newton, I. (au.), Cohen, B., Whitman, A. (trans.), *The Principia: Mathematical*
3 Khan, S. (2009). *Birthday Probability Problem*. YouTube: Khan Academy.

1.8 Cautions to New to Industry Technical Personnel

There are some cautions everyone should be aware of when in industry. Here are some actual cases that may help you learn from the errors of others.

1.8.1 The Wrong Frequency

A bearing failure problem kept occurring on an overhung pump similar to that shown in Figure 1.8.1. It was taken to an outside shop for bearing replace.

The opportunity was taken to use a newly acquired spectrum analysis measuring system to determine the natural frequency of the rotor assembly. The rotor was mounted as shown in Figure 1.8.1 with an accelerometer mounted on the shaft. The impeller was "bumped" with a soft mallet and the frequency analyzed. It indicated there was a resonance concern at the operating condition. The new to industry engineer reported this back to his mentor who didn't believe the results. He came to the shop took one look at the setup and said, "Your measuring the frequency of the table the pump is mounted on!" The pump was remounted on a heavy baseplate that resulted in the frequency being several times higher than

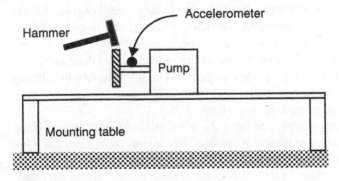

Figure 1.8.1 Overhung pump.

when on the table. This was very embarrassing and why it's mentioned here so others can learn. It would have been much worse if the results were in a technical paper.

1.8.2 Using the Incorrect Measuring Technique

Become a consulting after retiring from industry is a learning experience. On one job while reviewing the vibration data taken by site personnel, it was noticed that it was lacking the low-frequency harmonics that were expected based on an analysis that had been performed. Out on the equipment, you could feel these vibrations. A technique used when there is no measuring equipment present is to place a piece of paper on the machine and pull a line across with a pencil for a second or two, recording the time. Low frequencies will look like a sawtooth pattern, in the direction of the vibratory motion, so just count the peaks and divide by the seconds. The pencil and your body stay rigid and the paper on the equipment moves with the equipment. This will produce cycle/second, and while not very accurate it produces the peak-to-peak amplitude also. Crude but effective on low frequencies with fairly large amplitudes. This resulted in about 10 cycles/s. The analysis group said they would retest. Taking one of the team aside it was mentioned that the frequency range of the setup should be reviewed first to make sure the system was effective at such low frequencies. The next day at a review meeting the reanalyzed vibration data was presented to management and the low frequencies were present and agreed with the analysis. This is a good time for a consultant just to say "Good job!" It's always wise to discuss the results with the consultant before a meeting with management.

1.8.3 Never Under-Estimate the Value of Experienced People

Early in my career, I was witnessing the start-up of a large diesel engine and was standing on the west side of the engine. The Chief Engineer and Test Shop Superintendent were standing on the east side. The superintendent waved to me to come to their side. I had a much better view from where I was so waved them off. It was a little strange that the floor on my side was dark and their side looked like freshly painted gray. The engine started up, an oil line broke and sprayed oil on my front. I turned and ran and it sprayed oil on my back.

The next day the superintendent came to my desk and said, "Always stand behind the most experienced person at the test." Similar advice was given at a fire drill at an oil refinery. "Always follow the Machinists, they're running from the fire, the Operators are running toward it to shut things down." This wasn't such good advice since some of the machinists were also volunteer fire men.

1.8.4 Check and Double Check Your Design

Early in my career, I was given the job of designing an oil level measuring device for one of those large mining trucks used for carrying ore out of open-pit mines.

The sketch shown in Figure 1.8.2 is similar to one of the trucks and the driver, for scale reference, is about 6 ft tall. The company I was working for designed the motorized wheel the oil indicator was going on. It wasn't a complicated design but since there were several different versions many dimensions had to be verified.

After designing it, I had my colleagues follow the instructions to install it on a truck. It seemed acceptable so I took it to my manager. He gave me an airplane ticket and said, "Go to the ore mine out west and show them how to install it." When I got there I went to the Mine Supervisor, he looked at the instructions, chuckled and called in "Tiny" the welder who was going to install it. Tiny was 6 ft 6 in. and weighed about 300 lb. He drove me out to truck # 265 and I immediately knew what the chuckle was all about. The truck was parked in the normal 4 ft of mud and the top of the newly designed level was set at 3 ft. It would be 1 ft under the mud. Pretty difficult to pull out the level indicator to check the fluid level. The worst part was when Tiny asked, "Do you want me to get started or do you want to think about it for a while?"

Figure 1.8.2   Off-road mining truck.

This is where I learned, early in my career, to always understand all of the operating conditions on whatever you design.

1.8.5   Some Understand the Equipment Much Better Than You

An automated brake grinding machine produced the finish grind curvature on truck-type brake linings. It would periodically produce out-of-tolerance linings. The machine was fast and ground the shape in about 1 s after automatic fixturing. In order to observe the operation a video at 5,000 frames/s was taken. The key point is that the personnel responsible for operating, adjusting, and maintaining the equipment were invited to a meeting along with the plant manager and supervisors, to view the video.

In the middle of the session, the technicians and operators started talking to each other. It seems that they saw the problem in slow motion. An adjustment indexing arm would periodically work itself loose. It corrected itself when the next piece was clamped in place. This caused the product deviation. This was a random occurrence and happened to be caught during this video capture. This was fixed on several machines and solved the problem.

Those who are around equipment continuously usually understand it better than the manufacturer does.

1.8.6 Summary

It's important to learn lessons from others both good and bad. Only in this way can they be followed and not repeated.

1.9 A Method for Analyzing Catastrophic Type Failures

A portion of my consulting career involved investigating catastrophic type failures on machines, pressure vessels, and structures in industry. Causes and solutions were required quickly so the equipment could be safely restarted and production resumed.

While collecting and analyzing the damaged and mangled parts, my involvement has been to propose and test theories on the most probable causes. Some form of an analytical analysis is used based on the available data. The analytic methods used are usually taken from one of the Refs. [1–3]. I share this information with the investigating team and modify this as new information is obtained. This helps focus the team on all the confusing debris and input from interviews and helps in understanding what might have happened. Testing the theory in this way will either validate it or provide reasons to dismiss it. This is similar to what is done using the scientific method. In the scientific method, a theory, hypothesis, or best guess, on how something has occurred is proposed based on the data available. As additional data is gathered, such as metallurgical/material evidence, interviews or a mathematical analysis, it is tested to see if it agrees with the theory.

This approach is different than speculating on a possible cause. Speculating is saying what you think may have happened without this being verified against all of the available data. You are not trying to prove or disprove your thought.

It has been my experience that several events usually have to occur to result in a failure. Lack of or delayed maintenance, overloads due to exceeding operating or design specifications, faulty repairs, or modifications are typically the causes. Major design deficiencies from the manufacturer are rare if built to their specifications. A long history of successful operation before a failure of the type experienced would direct attention to one of these possibilities.

Consider theorizing on the wreck of a 2,000 HP gas engine–reciprocating compressor as shown in Figure 1.9.1 (1).

A sampling of the data assembled is shown in Table 1.9.1 to help in verifying the theory.

After reviewing the data shown in Table 1.9.1, which is only a small sample, the following theory was proposed.

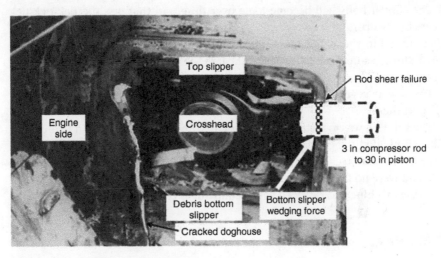

Figure 1.9.1   Wreck of doghouse section.

Table 1.9.1   Data to help verify theory.

Observation	Source
1. An experienced operator heard an unusual continuous tick-tick-tick and then a bang-bang just before the failure.	Interview
2. The compressor piston rod had sheared off due to wedging of the broken slipper.	Calculation performed along with metallurgical analysis. One million pound-force possible to do this.
3. The slipper bolts had all sheared off due to dynamic impacting.	Calculation performed along with metallurgical analysis. Loose hold-down bolt torque and shear probable.
4. The slipper bolts hadn't been torque checked for many years.	Maintenance records and interviews.
5. There were two bolts with low torque on the six other slippers out of 24 bolts.	Torque wench
6. The safety wire was still in place.	Observation
7. The slipper was cast iron and was broken into pieces and the 3-in. diameter piston rod had sheared. The nut was tight.	Observation and torque wrench
8. This type of failure had happened to the same type of unit at another division.	Interview with a colleague at a gas compression plant

The slipper bolts had become loose over time and that was the tick-tick-tick heard by the operator. The dynamic load impacted the body of the bolts and caused them to fail in shear. The loose slipper wedged itself against the piston rod and when the cross-head came forward it sheared the rod. This was the bang-bang the operator heard.

There were several other causes suggested by others, however, the data didn't support them.

The solution was to install dowels to secure the slippers in addition to the bolts. This eliminated the cyclic load on the bolts that loosened them. Periodic torque inspections were recommended as a maintenance item.

There were no failures of this type for 15 years after the modification at which time the machine was replaced with a centrifugal compressor.

References

1 Sofronas, A. (2006). *Analytical Troubleshooting of Process Machinery and Pressure Vessels: Including Real-World Case Studies.* Wiley.

2 Sofronas, A. (2012). *Case Histories in Vibration Analysis And Metal Fatigue.* Wiley.

3 Sofronas, A. (2016). *Survival Techniques for the Practicing Engineer.* Wiley.

2

Evaluating Failures and Designs

Someone's Statement of Facts May Have Been Obtained from Someone Else's Opinion

2.1 Twenty Rules to Remember

Many of these rules have been learned the hard way or by watching them happen to others. They are well worth remembering.

Out of all the work I've published, these 20 rules seem to have gotten the largest positive response from readers and seminar attendees. For that reason, they are stated here again and will be elaborated on in some of the later sections. A colleague provides me quite a compliment when he commented that they should be framed and placed in every practicing engineer's office.

While rules cannot replace common sense or a logical and methodical approach, they can help avoid embarrassing situations. Here are 20 rules that have been helpful in troubleshooting failures which every engineer or technician will eventually have to follow to be successful.

The rules have been developed for practicing engineers in the manufacturing and refining industry but should be useful to most engineers and prospective engineers.

Rule 1: Never Assume Anything
Making a statement like, "The new bearings are in the warehouse and will be there if these fail" is someone's opinion. They may not be there, may be corroded, damaged, or the wrong size. The only way you can be sure is to go out and see for yourself.

Rule 2: Follow the Data
The shaft failed due to a bending failure, because the bearing failed, because the oil system failed, because the maintenance schedule was extended, is following

Unique Methods for Analyzing Failures and Catastrophic Events: A Practical Guide for Engineers,
First Edition. Anthony Sofronas.

the data. A string of evidence much in like solving a crime is necessary in problem-solving. When trying to solve problems the person with the data will be the one who can solve the problem. Without data all one has is experience, speculation, or guessing all of which can result in the wrong answer if it does not support the data.

Rule 3: Don't Jump to a Cause

Most of us want to come up with the most likely cause immediately. It is usually based on our past experience that might not be valid for this failure. Contain yourself and don't do this and compile data first. This occurs most often when there is a large meeting and everyone is trying to provide input. Be careful when someone of importance or someone who should know does this. Without data, it can short circuit the problem-solving or troubleshooting effort and focus on only one cause when there may be many interactions.

Rule 4: Calculation Is Better than Speculation

A simple analysis is worth more than someone who tries to base the cause on past experiences. Many an argument in meetings has been solved by going up to the board and performing a simple calculation. It's hard to argue with this type of data. Remember engineering is performed using numbers and anything else is just an opinion.

Rule 5: Get Input from Others but Realize They May be Wrong

Most want to be helpful and provide input as to the cause, however, it may not be credible. When interviewing operators, machinists, and others, there are sometimes personal factors that enter into what people say about the cause. This is especially so when one person doesn't get along with another. You need to be aware of these conflicts when collecting data.

Rule 6: When You Have Conclusive Data Adhere to Your Principles

Safety issues are a good example. Your position may not be readily accepted by others because of budget, contract, or time constraints. Before taking a stand, it is important to have other senior technical people agree with you because it could affect your career.

Whenever there are critical decisions to be made, that's the time to be part of a team or form a team to make these types of decisions. You don't want yours to be the only name on a document. Engineering decisions are by necessity based on assumptions as all calculations have assumptions built into them.

Rule 7: Management Doesn't Want to Hear Bad News

Don't just discuss the failure and the problems it can cause. Present good options that can also be used at other plant locations to avoid similar failures. You will not be popular if you don't have solid methods to correct the problem. You may not need to select which is the preferred option but you should have the advantages and disadvantages of each. The meeting will be a success if one is chosen or if a next step is outlined.

Rule 8: Management Doesn't Like Wish Lists

Only present what is needed and not what you would like to have. Adhering to company standards or national codes is usually a wise approach. There were meetings where someone tried to tighten up specifications due to their experiences. The specifications were tighter than recognized national standards or codes and increased the project cost significantly. This didn't go well for the engineer and he was not asked to be part of future projects which was damaging to his career.

Rule 9: Management Doesn't Like Confusing Data

Keep technical jargon to a minimum and present the information as clear as possible with illustrations, photographs, models, and examples. Keep the presentations short and concise. All too often we are proud of the analytical analysis we have done and think everyone else will be too. Most of the time management just wants the results and what to do next. Details of the analysis are best left to the final report, a trade journal, or a book you write.

Rule 10: Management Doesn't Like Expensive Solutions

Only present one or two cost-effective solutions with options, costs, and timing. That is our responsibility as engineers. Present the options best for solving the problems even if the next step is more testing to gather additional data.

Rule 11: Admit When You are Wrong and Obtain Additional Data

This is most difficult to do but when other data contradicts yours, it must be done or you will look foolish. In this book, it is mentioned that it is a good idea to have the metallurgical results of a failure available before you present your mathematical analysis. Early in my career, I had done this in the reverse order and the failure mode was different than what the Materials Laboratory later determined. The laboratory results were correct and I had to correct my report. It was difficult and embarrassing to do but it had to be done.

Rule 12: Understand What Results You Are Looking For

The analysis was to determine why the rotor cracked and not to redesign the machine. Too often we get so involved in the analysis we forget to just solve the problem. This is especially true for very complex systems.

Rule 13: Look for the Simplest Explanation First

A mechanical engineer might see that a new drive belt was installed too tight and broke the shaft. Computer troubleshooters look to see if the devices are plugged in. Automotive experts make sure there is fuel in the tank. You can then proceed to the next simplest and least costly fix.

Rule 14: Look for Least Costly and Easiest Solution

You need to understand what caused the failure first. For example, if a drive belt was too tight, train the machinists the correct tightening procedure. Put a placard on the equipment with the procedure and a caution.

Rule 15: Analytical Results, a Test, or Metallurgical Results Should Agree

When the metallurgical analysis says it was a fatigue failure and your analysis says it was a sudden impact, someone is in error. They should both indicate the same failure mode. This was discussed in Rule 11 and shows what can happen if you don't have them agree.

Rule 16: Trust Your Intuition

When you feel something is wrong but can't prove it, it's time to do an analysis and get additional data. Your intuition is that little voice in your head that says that this doesn't seem right. All the wiring in your brain stores data and observations you have long forgotten but they are still locked away. So when a shaft looks too small in diameter or a motor looks too small to do the job, then you have unlocked a past experience or something you have read.

Rule 17: Utilize Your Trusted Colleagues to Confirm Your Approach

Talking with engineering and field colleagues has been the most useful method for finding the true cause of a problem. I usually go out of my way to watch how a job is done or an analysis was performed. After performing an analysis or a design, have someone review the critical ones.

Rule 18: Similar Failures Have Usually Happened Before

It's your job to survey your company and the literature for the cause of these types of failures and see if it is useful data for troubleshooting this failure. Most pieces of equipment are fairly generic and experience similar types of failures. A plant might have several hundred centrifugal pumps. Somewhere in the plant, someone has made a repair to prevent a failure, for example, hot alignments on certain types of pumps. It pays to be aware of what others have done.

Rule 19: Always Have Others Involved When Analyzing High Profile Failures

When safety, legal, or major production issues are involved, it's unwise to make critical decisions on your own. This is the time for a team approach so that nothing is missed and you have others involved to develop and implement the final solution.

Rule 20: Someone Usually Knows the Failure Cause

It has been my experience from interviewing engineers, operators, machinists, and technicians that several usually knew the true cause of a failure. A good interviewing procedure is therefore an important part of troubleshooting. For those that know the solution, give them the credit they deserve.

2.1.1 Summary

Reviewing these rules can keep one from making poor judgment calls.

2.2 How to Avoid Being Overwhelmed in a Failure Situation

When you're involved with a major failure and are looking at pieces scattered all over the area like a debris field, it's easy to become overwhelmed. This is especially true when the pressure is on because the unit is down, production has stopped, and management expects you to determine the causes and address them so they don't happen again. Updates several times a day are usually expected. This is becoming more frequent as experienced problem solvers are retiring from industry and less experienced personnel are selected to solve these types of problems.

Most of the failures we deal with are of the less serious type, meaning safety was not an issue. Usually, they are of the production limiting type and you may be expected to solve it on your own with input from others. Getting the unit back to full production safely, quickly, and reliably with the least expenditure is usually paramount.

I use to keep from being overwhelmed by recalling the following chart as a starting point and shared it with others involved in the investigation effort.

Figure 2.2.1 is pretty self-explanatory for machines and structures. It's a good way to start working a failure investigation when there's a group of people gathered

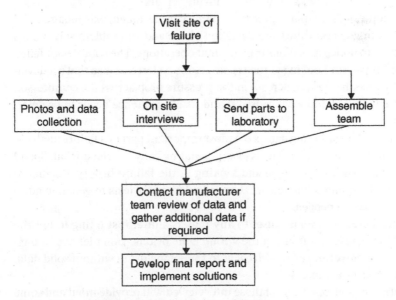

Figure 2.2.1 Typical approach to solving a failure.

in a room. Individuals can use the chart as a means for determining who will do what. For example, someone can be asked to contact the manufacturer and someone else follows the work the Materials Laboratory is performing. The names and timing of who will do what by when can be added to the chart.

One has to consider the type of problem that has to be solved. For example, a steam turbine that has lost some blades and had a major failure would be a candidate as long as there was no liability or safety issue. This method doesn't work well when handling cases involving management system issues where management may be a cause. Possibly the maintenance and inspection schedule had been modified to increase the production cycle. Fires, explosions, or fatalities also need a more formal review. Those types of problems are best solved with one of the current highly structured problem-solving methods taught in seminars. Such methods have an impartial team leader, independent from management, to direct the team's investigation into sensitive areas.

However, for many problems, it's useful to consider the following 10 points to help organize your thoughts:

1. Where is the deviation meaning what has changed? Was there a problem before the failure occurred, has a similar failure occurred on the equipment in the past, was a problem introduced during repairs?
   For example, let's assume a process compressor has failed, was repaired and failed again. Several things are deviations here and in problem-solving, it's important to notice deviations from normal operations. The compressor failed and was repaired. Looking at the repair area is a start because that's a deviation. Assume there is an increase in the pressure drop across the compressor. That's a deviation and what caused it and its consequence needs to be determined.
2. Remember that gathering data is the most important part of a failure analysis. An engineer without good data is just speculating. Anyone can do that. Spend time talking with field people and looking at the failure history. Document what you hear and what you see. Use calculations or tests to generate additional data when needed.
3. Usually, there are several causes to any major failure. Just trying to use the most likely cause can result in the wrong conclusions. Don't let one knowledgeable or powerful person drive you to their conclusion without solid data. They can be wrong too.
4. Validate your sources. Many of those interviewed will provide input and some of their data may be wrong. Always cross-check the data. When contacting the manufacturer of the equipment, they will usually provide a list of what the user probably did wrong. The user may not have done anything wrong and it might just be a poor design.

5. Don't get forced into providing an answer too quickly. It's more harmful to your career to provide a quick wrong answer than to take time and provide a correct one. Just remember when you listened to the misinformed newscasters who tried to provide immediate answers to the cause of an airplane crash and were wrong. Would you believe them next time?

6. Contact those you trust, meaning that senior technical people, experienced field hands, consultants, or the equipment manufacturer's technical representative. They usually have a good idea of what has occurred. Use your own judgment to filter such data. When things don't "feel right" to you, they probably aren't so trust your intuition.

7. You must solve the problem and make corrections and document your work. Have others included in reviewing what needs to be done and document it. Follow the job closely to make sure the plan is followed.

8. When presenting the causes to management, keep it simple, nontechnical and provide the best one or two solutions with cost and timing. Talk with the team members before such meetings so they understand what needs to be said. When this is not done, some may think this is an opportunity to provide their own opinion and not what was agreed on by the team.

9. Give credit to the participants. Nothing is ever done without the help of others such as inspection personnel, machinists, operators, and others so recognize their services when you can. It will be appreciated.

10. Completely document the effort and follow up on the success of the solution to the failure the team implemented. You may encounter a similar failure in the future.

2.2.1 Summary

As long as your approach is organized, logical, and well planned, the foggy nature of a failure slowly begins to clear and a solution will become evident.

2.3 Catastrophic Failures and the Human Factor

For this book, a catastrophic failure is defined as a failure where lives have been lost and human causes have been a factor.

Frequently I hear the question, "How could this ever happen? Don't Engineers know how to design things?" This always bothers me as there are usually several interconnected causes that lead to the final failure. Human action or lack of it is usually the primary cause and not design discrepancies. As mentioned in Ref. [1], "Complex systems almost always fail in complex ways." Several causes are listed here since they have occurred in the cases reviewed.

In this chapter, management means the decision makers in charge of the project. The workforce includes engineers, technicians, and tradespeople. The following causes are mentioned as major contributors in the cases reviewed at the end of this chapter. Being aware of these causes and using the suggested methods to address them is a way to prevent them from occurring.

1. Poor Communication Between Management and the Workforce.
   - This could be because management didn't understand the concern or it wasn't relayed to them in a form indicating its urgency. For example, "One of the workers has reported they have seen some cracking." Unless someone with the authority, knowledge, and the ability to assess the concern inspects the cracking and takes responsibility in addressing it, nothing may be done. In many cases, a knowledgeable tradesperson passes the critical safety information to a local supervisor. The supervisor may not pass it on to knowledgeable decision makers and the issue is never resolved.

2. Judgment Calls Without Adequate Tests or Calculations.
   - When a problem arises at a critical time in a project, management may not want to delay the project by performing the validating tests or calculations. Statements like, "It has happened before and it was OK," may be made. When the failure mechanism is not understood, this logic is inappropriate since if the first doesn't fail, what will happen the next time? The risk may be too high to ignore.

3. Calculations Are Made to Fit the Results Desired by the Management Team.
   - When something has failed, it can mean the calculations were inadequate. Trying to justify the adequacy of the design using the wrong analysis, techniques will provide the wrong answers.

4. The Workforce Fears Speaking Out about Their Concerns.
   - When one's job or advancement could be in jeopardy, there is usually a reluctance to pursue critical issues if told by management that it will be OK. Going to higher ups inside or outside the organization and the person feels, they will be considered a "whistleblower" [2] with all the complications that come with the title. However, it is important for personnel to speak up when they have sound evidence of a serious safety-related issue. When you are in a high enough position and say you will be documenting your safety findings, this usually gets action as it "puts the ball in management's court." When you are not in a senior position, having senior personnel or team members agree and support your position will reduce your vulnerability.

5. Not Wanting to Admit There Is a Critical Defect.
   - The person noticing the problem may have been part of the cause of the problem. Highlighting the problem will also highlight their involvement. Having

a well-thought-out plan or solution ready will be beneficial when you reveal the problem. Concealing it could be catastrophic.

6. Under Pressure to Get the Job Finished.
   - Bonuses may depend on getting the job done on a certain date and penalties may be imposed if the job is late. The contractors may feel that future jobs depend on finishing this job on schedule. Future funding for projects, bidding on future projects, credibility, and advancement of those involved may depend on the project's success. Pointing out that a failure will have far worse consequences and mentioning what they are may result in action.

7. Poor Review of Concern Areas.
   - When the job is relatively small, it may not be reviewed as closely. Also when there is a well-known and capable engineering team that did the design work, others may feel a detailed review is not necessary. Calculations may not be reviewed in detail or at all. A quick back-of-the-envelope calculation can indicate a problem. For example, the amount of damage done by the foam impacting the shuttle Columbia's wing was an elementary physics problem [3]. Presenting such information may result in action.

8. Areas of Concern By an Individual Are Not Addressed.
   - The person reporting the concern may not have done this in a forceful manner that conveys its critical nature to management. For example, a feeble, "I think there may be a concern," may not get any action. Also there may not be an adequate method to document areas of concern, except verbally. This means no one will be responsible for the concern area. Speak out forcibly if you see a catastrophic safety issue and recommend a process to document it as the organization may not have one.

Some notorious failures where many of the above causes were present:

1. Shuttle Columbia Disaster (2003) [1]
   - Insulating foam on the fuel tanks came loose and impacted the leading edge of the shuttle after launch. This problem had occurred before but was considered not to be a concern by management. It was listed as a maintenance concern rather than a safety concern and given a lower priority. Also, an improper impact analysis computer program was used to determine the effect of the foam impact. The analysis was developed for use in analyzing micrometeorite impacts on the heat-resisting tiles and not foam impacts.

2. FIU Bridge Collapse (2018) [4]
   - There was inadequate design and poor follow-up on addressing the cause of the large cracks noticed in the concrete during construction. This was a new, highly stressed, post-tensioned, concrete design. Cracking was considered

not to be a serious concern by the management team, but it was. Methods to correct the cracking only accelerated the collapse.

3. Hyatt Regency Walkway Collapse (1981) [5]
   - There was a miscommunication on the final design of walkway connections between contractors. The modified design doubled the supporting bolt loads with no redundancy present. Welding of the box beams had also been changed from the original design and was weaker. No review of the modifications had been done.

4. Bhopal Disaster (1984) [6]
   (a) Some final reports say the runaway reaction catastrophe was due to under maintenance of the plant and deterioration of safety practices. There appears to have been a series of neglect leading up to the explosion. Poor communications between plant management and workers were also cited. The Company's legal representation disagreed with the final report saying they had evidence that worker sabotage was the true cause.

5. Deepwater Horizon Oil Spill (2010) Off Louisiana Coast [7]
   - Miscommunication between contractors along with failure of the blow-out preventer and a poor cement casing job caused the explosion and subsequent spill. Serious warning signs, before the blowout, communicated by platform operations personnel were not addressed by the owners.

6. Shuttle Challenger Disaster (1986) [8, 9]
   - The "O" rings sealing the booster rockets failed due to the lack of flexibility in the cold temperatures at launch causing hot gas leakage and the explosion. The fundamental problem was a known design flaw. Management and the manufacturer had known about the design deficiency for nine years and failed to address it even though it had been communicated to them by engineers. The incorrect decision to launch in cold weather exacerbated the condition and engineer's protests of the decision were ignored. No escape module was incorporated for the crew because of management's false belief in the high reliability of the system. Even the final report by the investigating commission didn't quite understand the science involved [9]. For example, a safety factor of 3 was used that was strange. This was not true since if one "O" ring failed, the other would not hold and the safety factor would be zero.

7. Columbia Gas Transmission Pipeline Rupture [10]
   - A buried 20 in. diameter, 1,000 lb/in.^{2}, 45-year-old natural gas pipeline ruptured in a sparsely populated area without any injuries. Twenty feet of the pipeline was ejected from the ground and flew 40 ft. Fire damaged 22 acres and destroyed three houses with several others damage. The cost for repair was in excess of 8 million dollars.
     The NTSB report states that the failure was due to 70% corrosion of the pipe wall and was not a material or weld failure issue but due to the lack

of thorough periodic inspections. Management didn't address the alerts presented to them by monitoring equipment. The original pipeline design seemed adequate for the service.

2.3.1 Summary

The references presented detailed causes and solutions. The important fact is that such failures are many times due to human deficiencies and not just by erroneous engineering calculations or materials.

When critical safety issues are ignored, lives can be at risk, so it is important to speak out about your concerns. Try to do this using some of the techniques suggested in this chapter without resorting to being a "whistle-blower" as that can adversely affect your future career.

References

1 Columbia Accident Investigation Board, Volume I – VI, Government Printing Office, 2003.
2 OSHA, United States Department of Labor, https://www.whistleblowers.gov/
3 Sofronas, A. (2016). *Survival Techniques for the Practicing Engineer*, 152. Wiley.
4 Ayub, M. (2019). Investigation of March 15, 2018 Pedestrian Bridge Collapse at Florida International University, Miami, FL. OSHA.
5 Marshall, R.D. et al., Investigation of the Kansas City Hyatt Regency Walkways Collapse, NBS Building Science Series14-3, May, 1982.
6 Bloch, K. (2016). *Rethinking Bhopal*. Elsevier.
7 Report To The President, National Commission on the BP Deepwater Horizon Oil Spill and Offshore Drilling, January, 2011.
8 Space Shuttle Challenger Accident Investigation Board, Volume I – IX, Government Printing Office, 1986.
9 Feynman, R.P., Space Shuttle Challenger Accident Investigation Board, Volume 2, Appendix F - Personal Observations on Reliability of Shuttle.
10 N.T.S.B., Accident Report Columbia Gas Transmission Pipeline Rupture, NTSB/PAR - 14/01 PB 2014-103977.

2.4 The Importance of Alliances and Networking

You may be the best and brightest engineer your company has ever had; however, if you are unknown, this may all be in vain. I can't name many famous people who stayed to themselves, didn't publish their works or didn't have many who knew of them.

One can't sit in their office producing work and be successful in engineering. You are known by who you are and what you do.

I have found that just writing articles, technical papers, and books are not enough to become well known in engineering. It requires contact with others to develop the needed networks. Some of these contacts can be extremely helpful such as a chief engineer, design engineer, or field service engineer for an O.E.M. of equipment that has failed. With such alliances, others are available to help you with difficult problems or the need for data. This goes both ways as you will be able to help others too.

Some of the alliances that have proven helpful to me are:

- Engineers in O.E.M. plants and repair shops
- Equipment operators, machinists, supervisors, and managers
- Engineering colleagues
- Machinists and plant managers in repair shops
- Managers at plants I've performed work for
- Technical experts in companies I've worked for
- Technical experts in outside companies I've worked with
- Technical society leaders and members
- Equipment vendors who helped solve problems
- Participants in seminars I've presented or attended.

The list certainly isn't complete but the contacts have helped immensely in providing data and advice when solving difficult problems.

Throughout the book, you will see how some of these contacts have been made. The value of these people can be realized when they leave or retire from a company. Calling and hearing they are no longer with the company was always a letdown for me, like losing a good friend. Sometimes I could locate them again and the contact continued. When you cannot locate them, that particular level of expertise is now missing and can't be replaced nor is the friendship. Years of experience were needed for the advice the contact provided and it is doubtful others will have experienced and solved the same type problems. Hopefully, they have documented their work.

As a consultant and a company engineer, almost all of my work has been obtained from networking. Friends or others I have done work for pass my name on or recommend me for jobs. Very little of my consulting effort was obtained by someone just seeing my name in print.

2.4.1 Summary

Solving problems, presenting and documenting your results, and making valuable contacts are important in becoming well known in your profession.

2.5   Personal Checklists are Important

We usually feel we can remember important items when they are needed and may not write them down. When I took up flying, I realized my memory alone was not going to be good enough. It was self-preserving to see if the engine was developing full power and the instrumentation and controls were functioning properly before leaving the ground. So before departing, I ran through my written checklist of 15 items, mentioning each audibly while touching each gage or knob, much to my passenger's amusement. Becoming complacent about checklists can be dangerous.

Checklists are just as important during equipment start-ups. Whether you are starting up a 40,000-horsepower steam turbine or a small pump that is handling a dangerous product, your own personal checklist is a valuable tool.

Major projects have well thought out checklists called "punch lists." I'm not talking about those. I'm talking about your personal ones that help you from making those career-limiting mistakes (CLMs). A commercial pilot who forgets to review his checklist and doesn't put down the landing gear probably will survive, but his career might not.

While the list of items is small, they require a considerable amount of effort to ensure they are correct. These are usually items that have caused problems on similar equipment and required a shutdown or there was a safety issue.

Speak up early if something looks wrong. When it cannot be repaired before start-up, consider contingency plans. For example, if it's discovered late that there are hundreds of socket welds with poor penetration, consider a risk-based plan that will address the most serious faults first. For example, a high probability of a major pipe separation may need to be addressed before start-up, while the possibility of a low risk small nontoxic leak might be flagged for inspection or repair at the next downtime.

Many people are involved in major start-ups and engineers should be aware that high-level careers are enhanced for on time, below budget start-ups. Those who cause delays are not popular. You should have the support of others and very good reasons and documentation for causing a delay. Severe compromises on safety are good reasons.

The following sections in this chapter are condensed personal checklists used over the years.

2.6   Checklist for Vibration Analysis

Periodically, I receive questions from various worldwide locations on what might be wrong with a piece of machinery that was vibrating. It's impossible to accurately diagnose problems away from the machine system. For new to industry equipment engineers, the following advice was provided to prepare them for the investigation:

1. Are you sure the vibration measurement is accurate and the measuring system is performing correctly?
2. Are you sure the machine vibration is excessive? How do you know? Baseline data on what is normal vibration for the machine is always a useful tool. Vibration limits need to be understood as to why they were selected. Don't use vibration limits tighter than industry standards.
3. Was there always a problem with the machine even before the downtime? If it was present since installation, it may be a basic system problem such as a resonance problem, poor balance, an installation problem or just the wrong equipment selection for the service.
4. Did the problem start after a downtime and if so what was done to the machine during the downtime? For example, was a new rotor element installed? You can eliminate some possibilities if the original system was running well for many years. For example, any torsional vibration concerns were probably addressed in the systems design stage and changes in the system would have had to occur. Was the vibration monitoring equipment modified or was the machine realigned?
5. Was there some event that occurred which could have caused the vibration? Events such as power outages [1], operational changes, prolong surging in centrifugal compressors can result in rotor or seal problems, gear damage, or bearing distress. Processing difficulties and screw wear can cause failures in extruder systems.
6. Were the operating conditions changed such as gas properties, power, speeds, temperatures, or pressures? Compressors could be operating in a region where surge or some other phenomena is possible.
7. Has the system been modified, meaning new piping or vessels added or removed? Doing so could add thermal pipe strain and misalignment that could increase the vibration. A case was investigated where the system flow resistance was unexpectedly changed that brought a machine into an unstable surging range on the operating curve. In another system, dry gas seals were installed in a centrifugal compressor to replace the oil seals. The rotor developed rotor vibration issues because of the loss of the extra oil damping provided by the seals.
8. With gearbox unit vibration problems, have the gear mesh frequencies (GMFs) and their sidebands increased? This can mean the load on the gear teeth has increased. For compact high horsepower planetary gear and multi-reduction units, this can be complicated. Comparing the spectrum analysis data with the mesh frequencies may not always produce the correct answer unless the time domain is also reviewed. Time waveform analysis is the most useful and must

be done correctly. Know your limitations and use consultants who specialized in these types of problems when they are outside your area of expertise. Contacting consultants who do this type of work every day will usually pay for itself if you can predict one failure before teardown. It will allow you to have the spare parts on hand you require.

9. Remember to document everything significant. Similar vibration problems tend to reappear during one's career. They usually aren't quite the same but close enough so that historical data is useful.

10. Be familiar with the machinery in your facility. Walking around a piece of equipment daily can tell you a lot on how it is operating. When the bottom of your feet start tingling more than usual and strange new noises are heard, things may not be as they should be. Talk with the experienced operators on the unit as they will usually notice abnormalities with the machine.

11. Don't be too anxious to shut down an operating system and always have the decision to do so a group decision. The plant engineer, gearbox representative, vibration specialist, and management should all agree with the decision. The decision might be just to monitor and see if the vibration level increases with time. Once the machine is shutdown, no additional data on the cause can be obtained, other than by visual inspection of the internals. In a large, complex gearbox with many shafts, gears, and bearings this will be difficult. Starting-up without having determined the most probable causes would be most troubling.

12. With an operating system with high vibrations at a giving location, "mapping" of the system is usually recommended. This approach looks at various axial, horizontal, and vertical vibration amplitudes and their frequencies throughout the system [1]. These values are color coded on a plot plan and will ensure the critical location is identified. For example, worn splines on an extruder shaft or excessive wear of the screws may show up as a high vibration level on a bearing housing on the gearbox remote from this point. Without mapping all points, this wouldn't be known.

13. Question everything! When you are told the operation or processing hasn't changed, verify this yourself by reviewing the historical motor loading and vibration traces. For example, if the throughput meaning production remains the same but the product is harder to process in an extruder, this increases the loading and the GMF amplitude. Comparing the motor amp and vibration history may indicate this.

2.6.1 Summary

Complex systems and their problems can be overwhelming. With a systematic and logical approach and the help of others, the fogginess of a failure event will eventually become clear.

Reference

1 Sofronas, A. (2012). *Case Histories in Vibration Analysis and Metal Fatigue*, 152, 33. Wiley.

2.7 Checklist for New Piping System Installations

1. Is there a critical flange list and has it been followed?
2. Are all supports in contact with piping and shipping stops removed from bellows?
3. Do the welds look OK? A poor-looking weld usually will fail, especially socket welds.
4. Are hot bolting procedures available in the event of a gasket leak?
5. Do the piping stress isometrics make sense with what has been built? Are things where they are supposed to be?
6. Is everyone positioned safely during start-up? There is no need to have someone working on a piping that will become very hot.
7. Have you "walked the line" before start-up, scanning for items that just don't look right? For example, a pipe that goes nowhere or blinds still in place. It happens!
8. Has a hydrotest been done and have you reviewed the results on what was found and was it corrected?
9. Remember, if any of the piping is high-pressure steam, any steam leaks could be invisible and lethal. Have a way to verify there are no leaks, and make sure everyone is aware of the danger involved.
10. After start-up, have you "walked the line" checking for leaks and excessive vibration of the piping?

2.8 Checklists for Pumps and Compressors

1. Have you read the manufacture's start-up procedures on the motor, compressor or pump, and gear unit?
2. Has the alignment been checked and do the numbers make sense?
3. Has the inlet piping been cleaned?
4. Is the lube system OK and functioning? Check the schematics.
5. Is the motor, compressor, and gear unit oil drain back to the reservoir correct?
6. Are the seals piped up properly with the correct product used?
7. Is everyone positioned safely, meaning out of the way of leaking product, flying couplings or parts?

8. Has the vibration, temperature, and pressure instrumentation been checked out, and limits established such as when should you shut down?
9. Has the motor been checked uncoupled by electricians for correct rotation and no-load amps?
10. Steam turbine drivers should have their own checklist but it's essential that the overspeed has been checked and there is a method to check for those invisible high-pressure steam leaks that can be lethal.
11. Have the designated spare parts been checked and are they readily available? A spare rotor that doesn't fit or has a gravity sag isn't of much use. Neither are corroded, damaged, or the wrong bearings or seals. Just because a vendor says it's on the shelf doesn't make it so.

2.9   Understanding What the Failure Data Is Telling You

Before a failure analysis is performed, one important point should never be overlooked and that is that the failure data is trying to point to the cause. Usually, a broken part has a story to tell. Here are a few that have repeatedly been observed in failures the author has been involved with.

2.9.1   Gear Damage

The face of a gear as shown in Figure 2.9.1 can tell a lot on both the front face and back face. The unload gear face means the back face should be in pristine condition. When it is pitted or damaged, it is usually an indication of a torsional vibration problem or reversing a motor to unplug the polymer or elastomer from an extruder.

Heavy damage on the active side of the teeth with more on one end than the other can be an indication of misalignment.

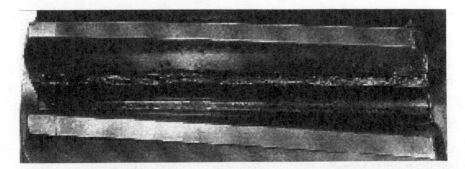

Figure 2.9.1   Gear face damage.

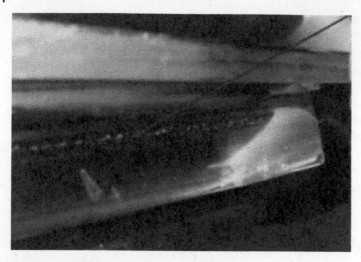

Figure 2.9.2 Dedendum pitting.

Figure 2.9.3 Contaminated lubrication.

Damage evenly across the face or at the dedendum as shown in Figures 2.9.1 and 2.9.2 can indicate various causes. Score marks as shown in Figure 2.9.3 can be traced to contaminated lubrication.

2.9.2 Shaft Failures

Cracks starting at keyways or fretted surfaces can be fatigue related and are usually the result of loose fits, rotating bending or vibratory torsional stresses. Sudden fractures that are perpendicular to the shaft axis surfaces or at 45° with the axis can

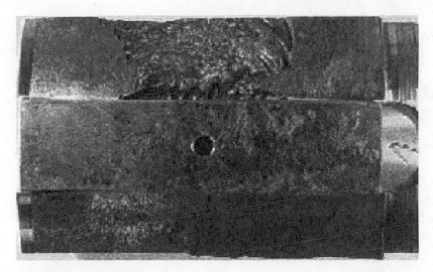

Figure 2.9.4 Fretted shaft at keyway.

be due to overloads. Shafts will usually fail in fatigue when the vibratory stress is above $\pm6,000\,lb/in.^2$.

Figure 2.9.4 shows a fretted shaft in the keyway area caught early. It would have eventually failed through the shaft.

2.9.3 Weld Failures

Welds will usually fail through their minimum longitudinal cross-section when the vibratory stress is more than $\pm3,000\,lb/in.^2$.

Figure 2.9.5 represents part of a disk dryer assembly in which two disks were held together using plug welds. A plug weld is a weld from one side only where the other side is inaccessible. This causes a stress riser at the gap as shown, where there was no weld material. A fatigue crack can start at this point if excessively stressed as shown in Figure 2.9.5.

2.9.4 Bolt Failures

The rule for bolts is to keep them tight. When they become loose, they will fatigue under vibratory loads.

Figure 2.9.6 shows a bolt that has undergone several sudden impact-type bending loads. This was one of several bolts in a large mixer that supported blades that were struck by accumulated product. The large chunks of this polymer product had conglomerated and wedged between vertical baffles and periodically fell onto the rotating mixer blades. The "beach marks" of the impacts are obvious.

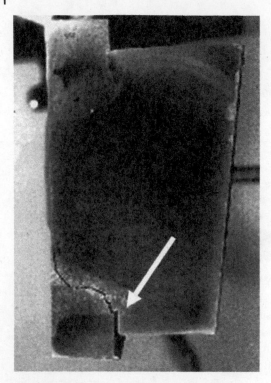

Figure 2.9.5   Crack growth of a plug weld.

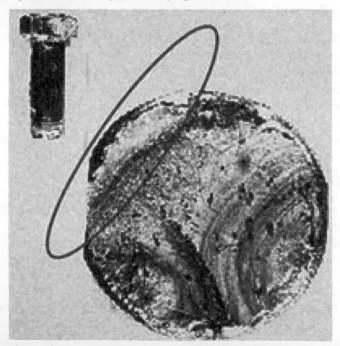

Figure 2.9.6   Impacted bolt bending load.

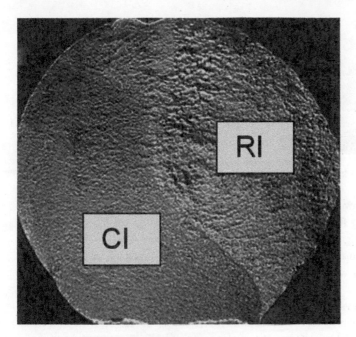

Figure 2.9.7 Bolt which failed in fatigue.

Figure 2.9.7 shows the fracture surface of a bolt that failed in fatigue. The surface is smooth and shiny in those regions that failed during crack initiation (CI) and growth. It is rough in those areas where it failed rapidly (RF).

2.9.5 Brittle Fracture Failures

Figure 2.9.8 represents the brittle fracture failure of a pressure vessel. This was a test vessel and the pressure was increased until a failure occurred at a weld

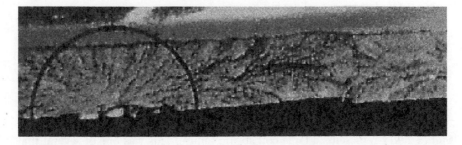

Figure 2.9.8 A brittle fracture failure.

Figure 2.9.9 Brittle fracture of a heat exchanger.

stress concentration. The tell-tale chevrons pointing to the fracture as well as the lines radiating from it are obvious. These are the types of fractures none of us want to investigate. Crack propagation approaches 5,000 ft/s. That's why it makes no sense to measure growing cracks in brittle material as it's too much of a gamble. It is wise to shut the equipment down and have them repaired immediately.

Figure 2.9.9 shows the brittle fracture of a heat exchanger during a hydrotest in a shop. The shop personnel didn't read the instructions and tested with cold water when it was supposed to be warm. It was also on a cold day. The material was never designed for cold service and broke off with a loud bang as it fell to the ground. The failure was due to a poorly made baffle plate weld.

2.9.6 Anti-Friction Bearing Failures

Most ball and roller bearing failures are caused by contaminated lubrication, no lubrication, or wrong lubrication. Bearing manufacturers provide detailed clues on what the damage observed means. Sometimes, it is said that the three-year L-10 life has been exceeded. However, this only means there is a 10% probability that the bearing will fail with the load prescribed and has a 90% chance of exceeding the three years. Most well-designed systems will have bearing that will last 10 years or longer before fatigue or spalling is a factor.

Figure 2.9.10 represents a bearing that saw repeated heavy impact loads. The loads spalled the races and cracked the rolling elements which resulted in a catastrophic failure. Repeated "bumping" of the motor to free a polymer plug was the cause.

2.9.7 Spring Failures

Figure 2.9.11 represents a large coil spring failure. This was one of several that were failing on a large vibratory conveyor. The failure was near the first small coil where it was bolted to the structure, and the stress was highest.

Figure 2.9.10   Thrust bearing impact failure.

Figure 2.9.11   Large 1-in. diameter coil spring failure.

It would be quite logical to think this spring might have been overloaded and cracks started on the corroded surface pits. Additional information came from a hardness check of the springs that indicated this batch of springs had been incorrectly heat treated.

2.9.8   Drilled Holes

Figure 2.9.12 shows the failure of a titanium connecting rod for a racing car. The fatigue failure is due to a faulty design. The small piston pin portion was designed

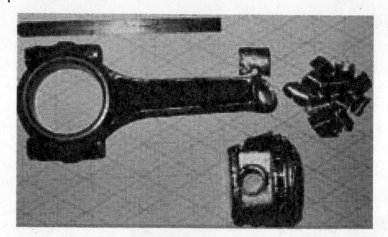

Figure 2.9.12 Connecting rod fatigue failure.

Figure 2.9.13 Secondary crack growth.

with an oil hole in it and was too thin. This caused a fatigue crack to grow from the overstressed oil hole.

Figure 2.9.13 shows a stop-drill hole in a stressed plastic piece but could just as easily have been metal. Stop-drill holes are small holes drilled at the end of a growing crack as a temporary repair until a permanent one can be made. The theory is that the drilled hole has a stress riser much less than the tip of a crack and therefore should halt the growing crack in the lower stress field. Figure 2.9.13 shows why this is only considered a temporary repair. At some point, this part was highly stressed again and the hole itself acted as a stress riser, causing a secondary crack to start from it.

2.9.9 Summary

It is imperative to save all relative failure parts after a catastrophic failure. The parts have a story to tell and can confirm or negate what an analytical analysis has determined.

2.10 Phantom Failures and Their Dilemma

While it would be nice to say all troubleshooting efforts have been successes, this wouldn't be true. Those who say they have always found the true cause of failures haven't tried to solve many problems.

There are those failures that haunt us called phantom failures because they are so elusive or so uncommon. Many times we make changes so the problem doesn't reoccur but really haven't found the true causes. This is troublesome because there is always the concern that it will reoccur.

Here are a few examples:

(1) A mixer/reactor vibrated excessively, however, when it was opened up and inspected, no cause was found.
(2) A pipe falls out of the pipe rack and ruptures for no apparent reason.
(3) A pipe fails from fatigue at a weld but there is no vibration in the system.
(4) One diaphragm in a steam turbine buckled from excessive force, but there were no apparent operating conditions that could have resulted in such a large force.

The following techniques have been used in the above failures:

(1) The mixer/reactor was instrumented for continuous velocity vibration recording to determine at what part of the batch process the vibration occurred. It was during the wash cycle and it was theorized that product had adhered to the vessel wall and was falling off periodically and being chopped by the rotating blades. This caused the vibration. The hot wash oil dissolved the product so that the evidence was gone when the teardown was performed. More frequent cleaning solved the problem.
(2) The sudden closure of a valve caused water hammer and the force knocked the pipe out of the rack. An analysis revealed this was possible. An alternate type check valve was installed by a valve specialty company.
(3) Two-phase flow occurred in the system during operation and the severe slugging cause this remote piping connected to a vessel to vibrate and fail in fatigue. Two-phase flow was not supposed to be possible in this system. The system was redesigned.
(4) An incorrect start-up procedure was thought to have been used, although no one admitted to this. Confined water instantaneously vaporized into steam and overpressurized the system. An analytical model revealed this was possible so the start-up procedures were modified. This solved the problem as no further failures occurred, but the true cause was never determined.

Thus, for the cases shown here and for other cases, the following approaches are sometimes successful:

- Instrument up the machine or system with vibration, strain gauges, torque, force, displacement, pressure, temperature, oil particle sampling or whatever is required and continuously monitor the results. The hope is to capture the next failure, if one occurs, and acquire the data needed to address it. Unfortunately, experience indicates that after a few weeks, if the problem hasn't reoccurred, the monitoring equipment is needed at another location, or is inoperative. It is usually right after this when the next failure occurs, thus no failure data is captured. So keep monitoring as long as practical.
- From the failure analysis data that has been collected or the analytical model that has been built, address as many of the potential causes as you can. This is sometimes called the "shotgun approach." It's not pretty but is better than not doing anything. If the failure occurs again, at least you have eliminated several possible causes. This is one of the major advantages of analytical modeling since many potential causes can be simulated on the computer without disturbing the operation of the unit.

Allowing even minor repeat failures by simply repairing because the cause cannot be determined can often escalated into more serious failures. It is therefore important to fully investigate all critical failures, even the phantom ones.

2.10.1 Summary

Either instrument up the system and try to obtain some data before another failure occurs or implement multiple solutions to the most probable causes identified.

2.11 Various Types of Equipment and Their Failure Loads

Details on some of these loads are discussed in various sections. This type of data is to alert the reader that there are methods that can be used to determine the failure loads. This doesn't imply that these are the actual loads you will experience. They are loads I have experienced when troubleshooting, based on analysis.

Agitators and Product Mixers:

- When the blades or paddles go in and out of the product and are not fully immersed, the mean torque can vary by 3× or more. This is similar to what occurs when a propeller goes in and out of the water. The motor speeds up and slows down, due to the load variation.
- The force on an agitator blade due to product impact, such as a chunk of agglomerated rubber in water can cause the blade load to instantaneously increase by 2× or more over the mean load.

Piping:

- Severe water hammer can increase the reaction loads on piping supports by 4× or more.
- Mixed flow conditions meaning gas with liquid slugs can increase reaction loads, especially if bending moments by 4× or more.
- Pipe spargers blowing gases into a liquid can cause fatigue failures due to vibration.

Pumps:

- Pumps operated on bypass to suction when not cooled can quickly overheat small volume pumps due to heat build-up.
- Cooling tower circulation pumps that lose power and have discharge check valves can develop high impact loads. This can damage the pumps due to backflow and the check valve slamming shut.

Extruder and Augers Driven By Motor via Gearbox:

- Normal torque fluctuations can be ±10% on the extruder shafts of the mean extruder torque in the gearbox on hot polymer and elastomers extrusion. An increase to 30% plus is possible with difficult products or heavy wear. This can cause torsional vibration problems to gearing.
- Repeated bumping, or reversing extruder screws to free plugging can cause cracks and fatigue failures on shafts, bearings, and gearing.

Centrifuge:

- Depends on the size, speed, and mounting structure. The unbalanced shaking forces due to products such as cold paraxylene clumps can be 4× the normal operational unbalance forces when nozzles plug.

Torsional Vibration Torques of Systems:

- The magnifier or increase of a fundamental resonance peak over the static forced amplitude can be up to 50×. Most systems were 10× or less.
- When the vibratory torque is equal or greater than the mean torque on a gear, it can undergo "hammering" of the mesh, meaning torque reversals which, if allowed to continue, can break gear teeth. Several sources suggest the vibratory torque should be 1/3 the mean torque at that speed.
- Vibratory torque greater than the mean torque can result in gear teeth separation and chatter at 2× GMF. The amplitude will be through the gear backlash and impact both sides of each tooth.
- The margin of a resonant frequency from an excitation frequency is usually taken as ±10%. This doesn't consider that the coupling spring constant provided by manufacturer can vary by ±20% that could change the natural frequency by 5%–10%.

Gear Box Tooth Load:

• This depends on what is being driven but could be 2× more than during smooth operating loads. When the vibratory torque is greater than the mean torque, impact pitting or gear chatter can occur. This can adversely influence the life of the gear face and pitting may be visible on the loaded face and unloaded face. With excessive pitting, the tooth can break off.

Reacceleration Loads:

• When power is briefly lost and reapplied to a large generator, the torque to the driven equipment, such as gears, may be 7× or more than the mean torque. This can develop cracks in shafts and gear teeth.

Shaft Failures:

• Excessive tension on a "V" belt can actually cause a cantilevered shaft to yield or increase bearing loads by 4× the manufacturer's recommended tension.

Synchronous Motor Start-Up Torque:

• The starting torque of a synchronous motor can have a cyclic torque of ±2× or more and will usually smooth out to less than 15% after full speed is reached. It depends on how fast it accelerates through the critical. It has many harmonics that can excite many system natural frequencies.

Pumping Pressure Fluctuations:

• The gas pressure fluctuations from positive displacement compressors can be ±20% of the mean pressure, while ±5% is common.

Surging Loads in Blower Air Systems:

• That loud "woosh" surge cycling can raise the blower static pressure by ±25%, while ±5% is common.

Vibrating Conveyors, Augers, and Other Fabricated Machine Fatigue Failures:

• When the cyclic stress on welds is ±5,000 lb/in.^{2} or higher, fatigue failures have occurred.
• Impact stresses of over 30,000 lb/in.^{2} can start cracks in fillet welds.

2.11.1 Summary

The data presented here can be used to show that a detailed analysis is necessary to help confirm and address a cause so it doesn't reoccur. They are just guides the author has experienced and shouldn't be relied on for design.

3

Mechanical Failures

Calculations Are Better Than Speculation

3.1 Preventing Crankshaft Failures in Large Reciprocating Engines

Crankshaft breakage is a rare event. While I have investigated the cause of several crankshaft failures during my career, most of the time was spent trying to prevent them from occurring.

The gas-processing and pipeline industries use many integral gas engine reciprocating compressors with crankshafts over 20-ft long. Always of concern was foundation and bearing damage. Either could cause the crankshaft to flex excessively with the possibility of a catastrophic fatigue fracture.

Figure 3.1.1 illustrates this exaggerated situation on a single-throw crankshaft. The overlap zone shown through the web is typically where fatigue failures occur. In this model, all web deflections are assumed to occur.

When the center-line of a crankshaft statically deflects web measurements show this. Analysis of this can indicate the source of the problem. This chapter shows how it's done.

A dial indicator or an electronic deflection instrument is placed at position (*A*) at bottom dead center (BDC) and set to zero. Web measurements are made by incremental turning of the crankshaft and measuring the web deflection from BDC to top dead center (TDC) at 90° increments. While the 90° and 270° horizontal readings are indicators too, the vertical readings are most revealing. The difference between the BDC and TDC reading is the total vertical web deflection. "Closing" of the throw means pushing the dial indicator plunger in as it is rotated from BDC to TDC and is regarded as negative; opening is considered positive. This is compared to the manufacturer's permissible limit which is typically ±0.18 mil/in. of

Unique Methods for Analyzing Failures and Catastrophic Events: A Practical Guide for Engineers, First Edition. Anthony Sofronas.

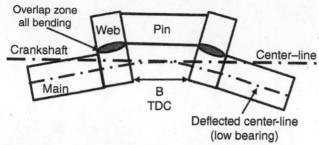

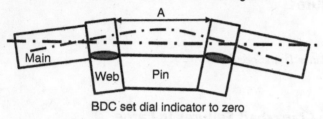

Figure 3.1.1   Distortion of a crankshaft throw.

stroke. Experience is required to interpret this value since some conditions can result in higher stresses than others, as will be shown.

It's also useful to relate the indicator readings to the crankshaft shape change it would cause. A traditional graphical model was used which converts the dial indicator readings into slopes and along with geometry calculates this change. The output from such a model is sometimes misunderstood. It doesn't consider the support of the bearings or how the throws are oriented. All throws are considered to be in the position shown in Figures 3.1.2 and 3.1.3. The crankshaft is considered unsupported and the vertical deflection is set to zero at the first crank throw. It is then allowed to develop the shape dictated by the dial readings, thus the strange curve shape.

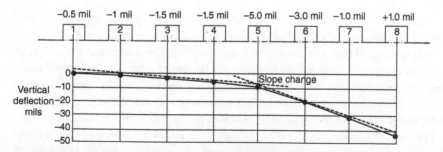

Figure 3.1.2   Crankshaft unsupported shape Clark TLA-8.

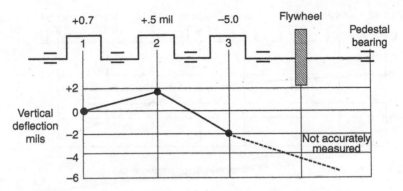

Figure 3.1.3 Cooper Bessemer GMV-6 with pedestal bearing.

The results of the model magnify the crankshaft distortion and along with the dial indicator measurements help make decisions and minimize the risk. Some examples will explain this.

Figure 3.1.2 illustrates the calculated shape from web measurements on a Clark TLA-8, 2,700-HP (HP, horsepower) integral gas engine compressor. This engine had main bearing distress to the right of crank throw 5. Notice how the slope changes after the excessive dial reading at throw 5.

While the dial reading of (−5) mils was over the manufacturer's maximum permissible of (−4) mils for this engine, it's effect is less severe. This is because the maximum power load on the rod occurs at TDC which tends to spread the webs apart so the alignment deflection stress is not increased. This type of information helps when management asks if the engine can be run for a couple of days so an orderly shutdown can be scheduled. The babbitt bearings were eventually converted to solid aluminum alloy main bearings to handle the uprated power of the engine.

Movement between the engine frame and foundation can cause grout wear. This occurred on a TLA-8 with a cracked foundation block and resulted in excessive dial indicator readings on throws 7 and 8. The deflection became negative in this region because the centerline had bent down. Wear is a relatively slow process so again a decision to delay immediate shutdown was made with minimum risk. Epoxy chocks were installed at all anchor bolt locations by a competent contractor. These could be selectively replaced when needed. This approach was used for 10 years until the engine was replaced with a centrifugal compressor. It was a cost-effective technique compared to replacing the foundation block.

A Cooper Bessemer GMV-6 integral gas engine compressor had an external pedestal bearing supporting a 3,500-lb flywheel. Alignment of the external pedestal bearing support was always a maintenance concern due to settling, lubrication problems along with temperature effects. Changes from positive

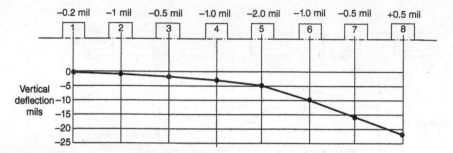

Figure 3.1.4 General electric FDM-16 diesel engine powering A ship.

to negative as shown in Figure 3.1.3 were quite concerning as it represents a complete stress reversal, much like a "kink" in a shaft. The pedestal bearing was eventually eliminated when a lightened flywheel was designed with approximately the same WR^2. No failures occurred with the new design which operated for 15 years at which time the engines were retired from service.

Shown in Figure 3.1.4 are the web deflections on a General Electric FDM-16, 3,100-HP diesel engine driving a reduction gearbox box for a ship main propulsion system. This was a new installation and it's in good alignment. Ships are special cases since they are flexible and the mounting of the engine to the ship is important as is the ships load distribution. Notice that the change in slope is not relevant here because the dial indicator readings are well within the manufacturer's allowable limit of ±4 mil.

There are other things I've learned after 15 years working on these types of engines.

- It's always wise to look at the dial indicator readings and discuss them with machinists who make these measurements often.
- Taking the dial indicator readings is hard work and requires very dedicated and thorough personnel.
- Take readings in mils or 1/100 mm, so you won't record the wrong number.
- More than 2 mil peak-to-peak motion between the frame and chocks was too much and eventually caused wear. Tightening down the anchor bolts never stopped the motion for long because of the lack of a bond. The chocks had to be replaced to stop the movement.
- A change between cold and hot conditions is normal but it's best to be expedient and plan to get all the readings within two hours after shutdown to remain close to operating conditions.
- A major wreck on the compressor side didn't seem to affect the power end crankshaft but probably could on less robust engines.
- Eliminating troublesome maintenance conditions, such as a pedestal bearing, poor foundation, or babbitt bearings with a better design, will usually be cost-effective.

Following good operating procedures and preventative maintenance programs can greatly reduce the probability that a crankshaft failure will occur.

3.1.1 Summary

Web deflection patterns can aid in determining the condition of an engine bearing and foundation condition. Alignment problems can result in major wrecks.

3.2 Structural Collapse of a Reinforced Concrete Bridge

While writing this book, a pedestrian walkway bridge under construction for Florida International University collapsed in March 2018, with lives lost and others injured. How can such a thing happen was my first response when watching the disturbing video newscasts of the failure occurring. There was considerable speculation as to what caused the problem. To satisfy my engineering curiosity, there was data available to have some idea of what may have occurred. Of course, a full investigation is underway that will hopefully provide a definite answer in a year or two, but I was interested in understanding it, the day it occurred, before the official results, on 15 March 2018. Newspaper and television sources showed the collapse of the bridge in slow motion. It was taken from a dash camera by someone about to pass under the walkway.

Figure 3.2.1 shows that the bridge, which was only partially completed, appears to fail at the points indicated. Puffs of smoke as if concrete was exploding were visible. It slid off the left side pillar, dropped to the ground and shattered. The right side stayed on its pillar.

Some data was gleaned from the Internet and news articles:

- This is a prestressed concrete structure, not a steel-welded structure. The strength is from pretensioned (PT) cables and rods in the concrete. It is a highly stressed complex design.

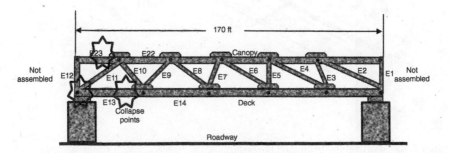

Figure 3.2.1 Sketch of bridge collapse points.

- It was designed to be a free-standing structure after it was moved into place.
- The bridge seemed to fail completely on one end but stayed supported on the other side as shown in Figure 3.2.1.
- The complete failure occurred in less than a second.

Pretensioning in concrete structures becomes very important for handling the 950-ton weight of the walkway, as do the truss joint connections into the concrete.

What is pre- and posttensioning (PT)? This is a well-tested method for handling the weak tensile properties of concrete when it has rebar in it. The beam is usually thick, with tension cables or rods in it. Essentially cables through the concrete are tightened to preload it in compression to overcome tensile loading. Posttensioning is done after installation to take care of concrete shrinkage and relaxation. To explain this a little better, consider Figure 3.2.2.

In this view, a concrete beam with a tensile cable load of (F) is shown. When the cable is stretched with a hydraulic jack, it exerts the tensile force (F) on the cable and a compression force on the beam (e11) when it is released. This puts the beam into compression and causes compressive stresses in the concrete and tensile stresses in the cable. This is much better than the original tensile stresses. Concrete has much less load capability with tensile stress than with compressive stress. Its tensile strength and shear strength are only about 10% of its compressive strength. Compressive stresses above 10,000 lb/in.^{2} in high-quality concrete can

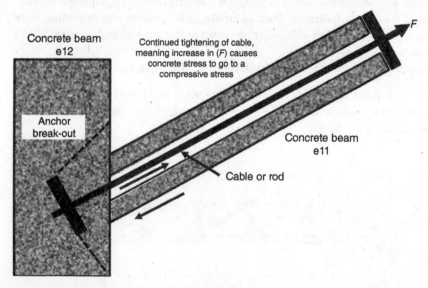

Figure 3.2.2 Prestressed beam and break-out.

result in failures too. In unreinforce concrete, shear cracks could be expected when shear stresses are above:

$$\tau_{ss} > 5^*(\text{concrete design compressive stress lb/in.}^2)^{1/2} \text{ lb/in.}^2$$

However, as is shown, there is also the possibility of a shear punch-out at the anchor point due to column load forces. These are complex connection points and where the cable tensioning is critical. This needs to be done very precisely or the concrete could fracture suddenly in an explosive and brittle manner.

A very rudimentary analysis which treats the structure as a truss with distributed load and pretensioning (PT) of the deck and canopy is analyzed. It considers the connection points as free to rotate, which they are not, but simplifies the analysis.

First let's reemphasize that the design calculations represent a difficult engineering challenge and require some complex software, complete data, and experience with the design of such structures. Most of the beams in the truss have been preloaded. All this analysis is intended to do is to see how high the loads and stresses were in the positioned structure at certain attachment points that had failed. Failure analysis calculations are different than code design calculations. The failure calculations are located at the failure points and utilize the failure data. Safety factors used in the design are irrelevant since it has failed.

The problem was simplified by considering the walkway as a structural truss as in Figure 3.2.3. A modified truss program was used to determine the forces on these elements. Only the loads in each link of the truss were analyzed from the scaled truss dimensions. A distributed load was developed from the walkway weight of 950 t. A PT of the deck cables of 8,000 kips total was used. For the canopy, 2,000 kips total from available design data. A PT load on beam E11 was 560 kips compression and E10 was 1,100 kips. A cross section of the structure and the PT cables are shown, performing a truss sensitivity analysis on the truss of Figure 3.2.3, resulted

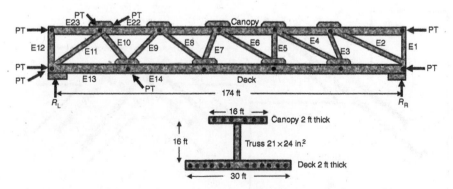

Figure 3.2.3 Loads in the walkway.

Table 3.2.1 Sensitivity study.

Link	e11 PT to 560 kips and e10 to 1,100 kips with deck and canopy also PT and dead load
10	700 kips
11	1,800 kips
12	500 kips

in the loads in all the links. The term (kips) means in thousands of pounds force. Only the selected links needed for further analysis are shown in Table 3.2.1.

All that can be said from this analysis is that the loads of the stand-alone PT structure are quite high and sensitive to the PT values used in the structure.

A complex connection exists which connects each truss beam to the deck or canopy. The failure pictures of the structure show that one possible failure point was at the connection E11 to E12 to E13. The other was at E11–E10–E22–E23.

Figure 3.2.4 shows the connections, shear zone estimations, and the imposed loads.

The push-pull out of the wedge area due to the compressive force of column E11 pushing and pulling on E12 and the deck is 2,200,000 lbs and the area in shear is estimated as $A_{shear} = 1,900$ in.^{2}. The punch-out shear stress on the concrete wedge area is about 1,200 lb/in.^{2}. For the canopy between E22 and E23, the punch-out/pull-out load is about 1,800,000 lb and the area in shear is about 1,700 in.^{2}, so the punch-out shear stress in the concrete is about 1,100 lb/in.^{2}.

These are pretty high nominal shear stresses for concrete with shear failures outside of the rebar especially with stress risers present. It is quite possible that once

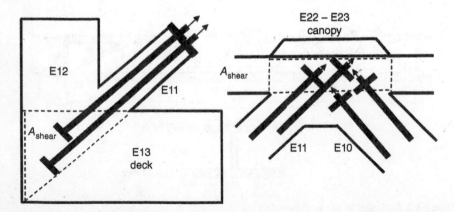

Figure 3.2.4 Anchor connection and break-out.

E11–E12 failed the walkway would buckle at the deck as shown in Figure 3.2.1 as there appears to be no redundancy.

3.2.1 Summary

The loads in the concrete structure appear to be quite high.

Cracks were noted in the concrete at E11–E12 at the time of the PT and before positioning into place. It is quite possible that the final design data was much different than the preliminary March 2018 data used here. This simple analysis allowed the construction to be understood better along with the complexity of the highly stressed design. This analysis did not determine the failure cause. Reference [1] is the final failure report issued June 2019 and has photos of the actual collapse. It stated structural design discrepancies in analyzing the loads and not addressing subsequent cracking problems noted on the structure. The failure zone in E11–E12–E13 was similar to where it was predicted to be in this analysis.

Two years after the failure, the causes were several and all debatable as is usually the case when litigation is involved. Preexisting cracks, erroneous design, and many other factors were cited and challenged. The abovementioned analysis that was performed a few days after the failure, for personal information only, seems to have captured a possible cause. The values may not be correct but the approach seems valid.

In fairness to the bridge designers and construction team, it was a very ambitious and complex design, and its failure was unfortunate and tragic. They were quite capable of designing a successful walkway but were using unfamiliar new techniques. Many of the points made in Section 2.3 occurred here.

Reference

1 Pedestrian Bridge Collapse 15 March 2018, *NTSB Rep. NTSB/HAR-19/20, PB2019-101363*, 10/22/2019.

3.3 Failure Analysis Computations Differ from Design

When equipment fails, it indicates where the weak point is and usually shows the cause of the failure. A designer of a new piece of equipment has to examine all potential weak points of the design. When design equations don't contribute to determining why something failed, they need to be modified. This will be explained with an example.

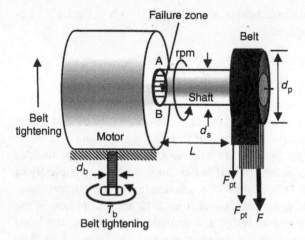

Figure 3.3.1 Motor shaft loading model.

Figure 3.3.1 shows a motor shaft with a power belt turning an expeller through a gear unit. Every time this motor shaft rotates, the shaft stays in a slightly deflected position due to the drive belt tension and pretension. The bending stress is cyclic and every circumferential point in the radius goes from tension at (A) to compression at (B) as the shaft revolves. This shaft has been designed for infinite cycle life by the motor manufacture for its intended application; however, the shaft failed.

Unfortunately, this type of failure has occurred several times during my consulting career in several different industries.

Modeling the manufacturer's analysis method by including factors that could cause failures help determine causes thus suggesting solutions.

Here are equations the manufacture might use in the design.

The torque in the shaft is:

$$T = 63{,}000{*}HP/rpm \ \ in. - lb$$

With a pulley diameter (d_p), this produces a force on the shaft (F).

$$F = T/(d_p/2) \ lb$$

The additional belt tension force (F_{bt}) is applied by tightening the belts:

$$F_{total} = F + 2{*}F_{bt} \ lb$$

An approximate value used by some manufacturers for belt tension is $F_{bt} = 0.2{*}F$.

The correct manner to tighten the belt is by measuring belt center span deflection with a force gage. Bolts are tightened to develop this static force (F_{bt}) by stretching the belt. The design equations can be modified to approximate this.

The moment on the end of the shaft is:

$$M = F_{total}{}^* L \text{ in.lb}$$

At full power, the shaft bending stress (σ_b) and shear stress (σ_s) are:

$$\sigma_b = 32^* k^* M / \left(\pi d_s^3 \right) \text{ lb/in.}^2$$

$$\sigma_s = 16^* k^* T / \left(\pi d_s^3 \right) \text{ lb/in.}^2$$

These can be combined as:

$$\sigma_{equiv} = \left(\sigma_b^2 + 3^* \sigma_s^2 \right)^{1/2} \text{ lb/in.}^2$$

This is straightforward for a new design when the manufacturer has control of all of the parameters. Usually a conservative stress concentration factor $k = 3$ is used for a well-designed shaft radius and keyway.

A conservative value for the shaft material endurance limit for a polished specimen is also used by considering surface finish, size, type load application, and other factors. This motor was 75 HP, 1,200 rpm, $d_s = 2.875$ in., $d_p = 10$ in., $L = 7$ in., tensile strength of steel 125,000 lb/in.^{2}, yield strength 100,000 lb/in.^{2}, shear strength 50,000 lb/in.^{2}, and a corrected endurance limit in bending of 30,000 lb/in.^{2} at 1^*10^6 cycles.

These values result in a total design stress $\sigma_{equiv} = 11,000$ lb/in.^{2} in the radius and keyway, which are well below the shear and yield strength. The cyclic portion which is the bending stress (σ_b) in the radius, and keyway is $\sigma_b = \pm10,000$ lb/in.^{2}. This is also below the corrected endurance limit, so the fatigue life should be unlimited.

This is a well-designed shaft, which was in service under the same processing conditions for five years and shouldn't have failed in shear, yielding, or fatigue. So why did it fail?

The failure zone in the shaft radius at (L) shown in Figure 3.3.2 supplies additional information.

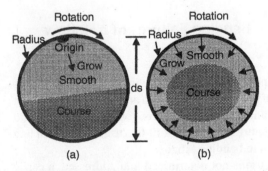

Figure 3.3.2 Shaft fatigue failure zone.

The fatigue crack had propagated halfway through the shaft by a rotating-bending type fatigue, which is shown in Figure 3.3.2a. This is indicative of a heavy cyclic stress. There was a defect in the radius or the failure would have appeared more like the rotating-bending of Figure 3.3.2b. This is because in Figure 3.3.2b the cyclic stress is even around the radius and the crack grows inward as shown by the arrows as the shaft rotates.

The load required to fail the shaft wasn't caused by a sudden impact since the failure zone is smooth indicating many cycles to failure. One day represents about $2*10^6$ cycles and beach-mark spacing, meaning opening and closing of the crack, would be about $1*10^{-6}$ in./cycle and would appear smooth with a visual examination.

A key piece of additional information obtained from field personnel was that a new drive belt had been installed the previous day.

Overtightening of the belt tensioning bolts when the motor was installed, without using correct measuring techniques, could increase the bending stress in the radius. For example, tightening the bolts for belt pretensioning by a factor of 10 was verified as something an inexperienced person could easily do. This would raise the cyclic bending stress from 10,000 to 36,000 lb/in.^{2} or higher and could fail the shaft in fatigue in less than a day, as was observed.

3.3.1 Summary

Design equations may not explain a failure. Adding factors that can explain the failure may be necessary. By adding belt tensioning (F_{bt}) to the basic design equations, the effect of over-tensioning could be evaluated. With training and a placard on the motor, a reoccurrence of this type of failure was unlikely at this facility.

Similar modifications to other design equations can help when analyzing failures and providing corrective actions.

3.4 Crack Growth and the Bending Failures of a Hollow Shaft

Troubleshooting shafting failures plays an important role in industry. Many types of machines can have these types of failures and have been analyzed by the Author. Extruder shafts, auger shafts, turbine shafts, crankshafts, rolling mill shafts, and many others have broken in half. Figure 3.4.1 shows a 10-ft extruder shaft that had suffered mid-barrel wear and failed in bending fatigue.

The fear always is that if the cause is not determined and addressed, a costly failure may occur again. There are many reasons for shaft failures but rotating

Figure 3.4.1 Typical extruder shaft bending fatigue.

bending type failures frequently are the cause and start from a defect of some type. A metallurgical report may mention this but probably won't tell you why it occurred or what to do to alleviate the cause. This case analyzes such failures.

The growth of a crack is usually a fatigue phenomenon. Most metals have cracks especially if there are welds. Even grain boundaries in the metal itself are a type of crack. Cyclic tensile stresses can open the crack and if these stresses are high enough these cracks will grow each cycle they are opened. There is a limit at which the stress field is not high enough for the crack to propagate. This means it will not grow in a ductile material when the cyclic stress is below this value. While debatable, from failures witnessed, this seems to be around $\pm 2,000$ lb/in.^{2}.

Traditional fatigue analysis approaches such as the Modified Goodman Diagram are used to see if designs are adequate. Since fracture mechanics can determine the stability of cracks in old pressure vessels, it was realized that crack growth calculations could be used on ductile materials also. Traditional fatigue methods aren't sufficient when there is a sizable preexisting crack. These methods usually determine how long it takes to develop a crack. Since a crack already exists crack growth calculations can be a very useful tool. So let's see how it can be used.

Consider Figure 3.4.2 which shows a hollow auger shaft with a sag. The sag could be due to the gravity weight of a large rotor between bearings or a horizontal mixer shaft with a weight or pressure load on it. With every revolution of the shaft, Point A goes through a tensile stress and a compressive stress when it gets to Point B in this sag position. A sag is like holding a rubber hose and turning the ends. The hose stays in the sag position as it is rotated and the tensile stress will tend to grow the crack. Now if the shaft were in a permanently deformed and bowed condition such as occurs with a thermal bow, which is not shown, for every revolution of the

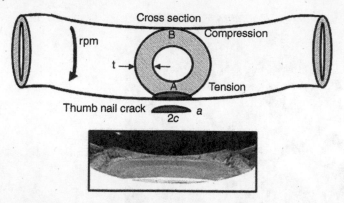

Figure 3.4.2 Sagging shaft with thumb nail crack.

shaft Point A will be a constant tensile stress even at Point B. It is therefore not a cyclic stress, so a crack will not see the cyclic stress and won't grow.

Figure 3.4.2 also shows a crack in a shaft with a sag. The analysis of crack growth is based on fracture mechanics, but since the discussion will be on ductile materials, the equations are simplified. Shown is what is known as a thumbnail crack because of its shape. The depth "a" and width "$2c$" define the crack and are the typical nomenclature used in fracture mechanics.

The approach will be to assume a small initial crack size and calculate the time it will take for the surface crack to breakthrough the wall thickness (t). When this happens, the shaft will have lost much of its strength and a total failure will soon occur.

For this analysis, the initial crack depth (a) will be assumed as $a_o = 0.01*t$ and the final at breakthrough $a_f = t$.

The life equation for stainless steel is obtained by integrating the Paris Equation (1) and after making the substitutions is:

$$N = [8.3*10^8/\Delta\sigma^{3.25}]*[1/(0.01\ t)^{0.625} - 1/(t)^{0.625}]\ \text{cycles}$$

Here $\Delta\sigma$ in ksi is simply the cyclic tensile stress opening the crack. It will need to be determined by analysis or testing.

The life to reach this number of cycles when the shaft rpm is known is:

Life in years $= N/(525,600*\text{rpm})$

3.4.1 An In-Service Failure Example

A failure occurred on a welded shaft made of a 6 in. diameter pipe with a wall thickness of 0.55 in., rotating at 40 rpm. The shaft has a sag due to its weight and the tensile stress using beam theory with a distributed load is calculated as $\Delta\sigma = \pm8$ ksi.

Proceeding through the analysis results in a life of 1.12 years. Since this was a mixer, additional mid-span support bearings were required.

The thumbnail crack assumed in the analysis is $2c/a = 6$ which is fairly typical when looking at failures. This means the initial surface crack for this case was 0.0165 in. and the final surface crack at breakthrough was 1.65 in. The stress would be higher because of the loss of the cross section, further accelerating the crack growth. A larger initial crack would produce a shorter life.

As a sensitivity analysis, assume the cyclic stress is reduced to $\Delta\sigma = 4$ ksi with the same initial crack. For this case, the life is 10.6 years and so a mid-bearing will certainly be a good solution.

3.4.2 The Assumptions and Comparisons

Like all simplistic models of crack growth, some justification due to actual experience is usually warranted. This type of analysis would never be used for brittle fracture studies but is useful when it has predicted ductile crack growth with some success. So from previous case histories, Table 3.4.1 has been established:

Again the actual and predicted values vary considerably with historical data; however, when no other data is available, sometimes it's enough to make an educated decision.

For example, when a crack was noted on a gear tooth root and management wanted to know if it could be run until the next planned downtime in a week an analysis was conducted. With a predicted life of less than eight hours, it couldn't and the risk was explained to the Owners. If the tooth broke off, it could fall into the mesh and cause a major wreck. The repair would take a week rather than a day. This educated guess with some analysis is better than saying it's OK and hoping for the best. The risk just isn't worth the reward. There was no need to explain the analysis and just mentioning one had been done was enough to have the gears

Table 3.4.1 Various crack growth cases.

Case	Shaft (rpm or cpm)	Cyclic stress (ksi)	Initial/final crack size (in.)	Life years actual/predicted
1. Extruder shaft (wear bending)	120 rpm	3	0.01/5.0	10.0/7.0
2. Dryer disk bending crack (measured)	10 rpm	6	1.0/2.0	0.4/0.2
3. Cracked gear tooth	650 rpm	16	0.1/0.4	Short/0.001
4. Vibrating conveyor welds	500 cpm	3	0.01/2.0	0.8/1.5
5. Vibrating conveyor welds repaired	500 cpm	1	0.01/2.0	+10.0 still going/54.0

changed out. Many times management is willing to take the risk for production reasons. This is fine as you would have presented the science as engineers should and their decision is being made for other reasons. As they say, "The ball is now in their court."

3.4.3 Summary

Crack growth calculations can be used on machinery on ductile materials. The time for a crack to grow to a failure length can be estimated. This is sometimes used to see if the equipment can be run until a planned shutdown when repairs can be made.

Reference

1 Sofronas, A. (2006). *Analytical Troubleshooting of Process Machinery and Pressure Vessels*, 340. Wiley.

3.5 Why Did a Small Piece of Foam Cause the Shuttle Columbia to Crash?

This is in this book because it affected everyone so much at the time. It was the catastrophic failure of the shuttle Columbia in 2003 due to an insulating foam failure.

At the time of the accident, before any investigation was done and the Columbia and crew were still in orbit, some news agencies were saying that there did not appear to be any damage. One report stated that the piece of foam that broke off and impacted the wing was like a styrofoam cooler mounted on top of your car. The report said, "Let's say it came loose and hit the car behind you. All it would do is break up into little pieces!". This view was strengthened since blurry telephotographic imagery after launch saw the impact and a shower of possible debris at the impact point [1].

In the case of the Columbia when the insulating foam came loose from the external fuel tank after it was launched, a large piece was known to have impacted the wing. Figure 3.5.1 is a simplistic sketch of this. Now to a mechanical engineer saying it would just break up with only a few pounds of force at the speed it was traveling sounds absolutely absurd. A pilot friend said he had hit a duck on the leading edge of his private aircraft wing at 150 mi/h. It penetrated one foot into the aluminum wing skin. For these reasons, the following simple calculations were performed to satisfy my curiosity.

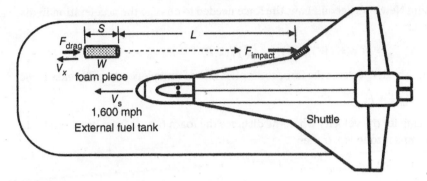

Figure 3.5.1 Impact of foam on shuttle.

Like all engineering problems, one gathers data, makes a sketch of what's happening, and then performs the necessary calculations. For this problem, it's just using Newton's Laws and calculating the deceleration of the foam piece before it impacts the wing. Videos were available that showed the foam coming loose and from these videos the approximate size could be determined. It was stated as "being bigger than the size and weight of a loaf of bread." So with a sketch of the shuttle and only the meager data available, some calculations were performed. All that was desired was a rough approximation. After all, some were saying that the impact force was only a few pounds, merely a styrofoam cooler breaking up at 60 mi/h.

The following are the calculations performed while the Columbia was still in orbit. At that time, I was waiting to see similar calculations on the NASA site but never saw them.

In these calculations, it is assumed that the foam piece hit the wing and the foam crushed its full length before it broke apart but didn't penetrate the leading edge. It was surprising there wasn't a simple analysis like this shown in the final report [1].

The first task was to understand with what velocity the foam hit the leading edge. Since it was important here's a practical description. As soon as the foam breaks off the external fuel tank, its speed is the same as the shuttle. With no air resistance, it would stay at approximately the shuttle speed and not impact the wing. However, as soon as it gets into the airstream at 65,000 ft and the air density is one-tenth that at sea level, the foam decelerates due to the drag force, and its velocity slows down. The shuttle because of its constant speed then runs into the foam with an impact speed of the shuttle minus the foam speed.

Here's the approach used to determine the impact on the shuttle wing.

The drag force at 65,000 ft decelerating the foam will be considered as a constant force on the foam:

$$F_{drag} = \tfrac{1}{2}{}^{*}C_d{}^{*}\rho/g^{*}A^{*}V_s^2$$

Using Newton's Second Law, the force needed to change the acceleration in distance L:

$$F_{accel} = m^*a = (W/g)^* \left(V_x^2 - V_s^2\right)/(2^*L)$$

Since the drag force is the force to decelerate the foam piece when it breaks loose:

$$F_{drag} = F_{accel}$$

Solving for the velocity when the distance the foam travels from break-off to the wing impact zone is L ft then:

$$V_x = \left(V_s^2 - F_{drag}{}^*32.2^*2^*L/W\right)^{1/2}$$

Here A is the drag area in ft^2, V_s is the shuttle speed in ft/s, W is the weight of the foam lb., and V_x in ft/s is the speed of the foam before it impacts the wing.

With $A = 1$ ft^2, $V_s = 2,400$ ft/s, $W = 1$ lb$_f$, $L = 80$ ft, $C_d = 1.0$, $\rho = 0.005$ lb/ft^3.

The impact velocity $V_{impact} = V_s - V_x = 540$ ft/s or 370 mi/h.

The following assumption that Δt is the time the foam deforms when it impacts the wing, $s = 1.5$ ft distance. Notice no consideration is given to the foam penetrating the wing. In this analysis, the wing is considered rigid.

$$\Delta t = s/V_{impact} = 1.5/(530) = 0.0028 \text{ s}$$

$$\begin{aligned} F_{impact} &= [W/g]\,(V_{impact})/\Delta t \\ &= [1/32.2]\,(540)/0.0028 = 6,100 \text{ lb} \end{aligned}$$

So with this, we see that an approximation of the force with which the 1 lb piece of foam would impact the wing was 6,100 lb or 3 tons. There's no need to do a stress analysis with this large of a force as it's quite evident that it will probably do a lot of damage on the thin carbon fiber wing. It's like dropping a truck on the wing. The force will be less because the following were not considered in the analysis:

- The foam will break up when it impacts distributing energy.
- The foam will penetrate the wing, and so the full force will not be realized.
- The foam will partially glance off the wing reducing the contact force.
- The foam may do all of the above.
- Details on many other parameters were not available or estimated.

As a comparison, using the methods just described, the impact force of a styrofoam cooler falling off a car and hitting one behind it going 60 mi/h and with sea-level air density, the force would be about 20 lb.

What it took to prove to the organization that there was a problem with the foam impact was a well-publicized test. An independent contractor fired a 1.7-lb piece of foam from a nitrogen-pressurized cannon at 770 ft/s at a partial shuttle wing. It produced a force of 4,500 lb on the wing before penetrating it and putting a hole in it about 16 in. in diameter.

3.5.1 Summary

The weight of the piece of styrofoam was small, but the velocity it hit the wing with was very high. The less than 2-lb piece of styrofoam impacted with over two tons for force causing a hole in the wing. Super-heated gas entered this hole destroying the shuttle.

Reference

1 (2003). *Columbia Accident Investigation Board*, Vol. 1, 248. National Aeronautics and Space Administration, Government Printing Office.

3.6 Can the Aircraft Cowling Contain a Broken Turbine Blade?

This question came up because a Southwest Airlines aircraft with a CFM 56-7B engine had just lost a fan blade, and it destroyed the engine. There was one fatality. Was there any possibility that the cowling (t_h) could capture and contain a blade?

In this analysis, the blade departs tangentially (V_t), because it is directed that way by the unbroken adjacent rotating blade and force (F) as shown in Figure 3.6.1.

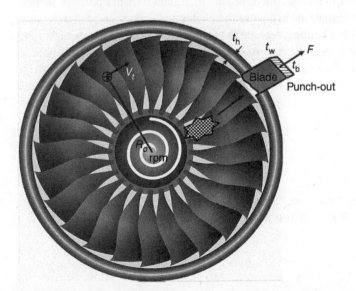

Figure 3.6.1 Similar CFM56-7B compressor fan blade broken.

The same analysis technique as was used in Section 3.12 will be used here; however, there are many unknowns in this type failure. The adjacent blades can also push the broken blade into the cowling. There may be many broken blades jammed between the cowling and the disk. There will be much heat and friction [2]. With all that in mind, all that can be said is that if the force of the blade punching-out the cowling fails it, the other factors will just make it worse.

The shear strength of a titanium containment band is $\tau = 95,000$ lb/in.^{2}, rpm = 5,175, $R_o = 15$ in., $t_w = 0.25$ in., $t_b = 2.0$, $W = 6$ lb, $\varphi = 0.5$.

A titanium-band thickness of $t_h = 0.25$ in. was assumed, but because of the complicated design, it was unverified.

The calculated value for $\delta = 0.80$ in. and since the band is only $t_h = 0.25$ in. thick, it won't contain the blade, using this analysis technique.

Reviewing failure photographs [1] seems to indicate that the failure didn't occur in the containment ring region. All damage was forward of the blades and the cowling broke away in a composite material section. The blade probably was thrown forward or the severe unbalance vibration and resonance might have caused the forward cowling fiberglass material to have failed.

3.6.1 Summary

The cowling assembly cannot be relied on to contain a broken blade as the unbalanced force is large and can result in other damage as seems to have occurred in this accident. Recent new designs, as seen in reference [2], can contain large blades which tests indicate can be contained. Engine manufacturers use periodic inspections and life cycle limitations to ensure the blades and disks won't fail in fatigue. A cycle is a take-off and landing. As long as periodic inspections are adhered to and engine rebuilds are done at the recommend life cycles, the engines should be safe. However, there are other unknowns besides fatigue. Hail impacting, bird strike ingestion, run-way debris, and poor pilot operating procedures can cause other damaging loads that can start fatigue cracks. Poor repair procedures or defective parts are always a concern.

References

1 Wikipedia. (2019). Southwest Airlines Flight 1380.
2 YouTube. (2020). Blade Off Test A380, youtube.com/watch?v=j973645y5AA.

3.7 Why Did My Car Windshield Break from a Very Small Stone?

A 1/2 in. (0.04 ft) diameter (*d*) stone falls off a truck traveling at 60 mph (88 fps), and you are following at the same speed. When it hits the ground and bounces up to your windshield height, it hits with about the same velocity your car is going. Assume the stones weighing $W = 0.004$ lb totally crush on your windshield on impact.

Equating kinetic to potential energy as one form of energy is converted into another form:

$$1/2(W/g) V^2 = F^*d$$

Solving for *F* of impact is:

$$F = (1/2)^*(0.004)/(32.2)^*(88^2)/0.04 = 12 \, \text{lb}$$

This doesn't sound like much of a force, but if there's a sharp corner on the stone meaning a small contact area the stress can be high. The stone made a 1/16 in. pit on the windshield, so assume a point that size. A crack had started from it.

$\sigma = F/A = 12/(0.785^*(1/16)^2) = 3{,}900 \, \text{lb/in.}^2$ which is in the glass fracture range.

The idea that very small objects cause damage at high speeds has always been obvious, but an analysis can quantify this.

3.7.1 Summary

The small stone has sharp corners and the light impact force causes the brittle glass to stress and fracture.

3.8 Momentum or Why a Car Is Harder to Push and Then Easier When Rolling

Sometimes you have to push your car by yourself. When you first start to push, it takes quite a bit of force to get it rolling. Once it starts rolling it gets easier. However, if you have given it a relatively fast push, it may roll on its own. It may even go a little way up a hill with no help from you. Why is this? Figure 3.8.1 shows the situation.

Several things are happening here. There is the force needed to overcome accelerating the weight of the car from rest and also the force to overcome the rolling

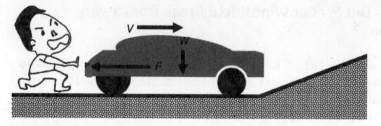

Figure 3.8.1 Pushing a car.

friction of the tires. There is then the additional force required to push the car up the grade due to the weight of the car and gravity trying to roll it back.

First, the force required to start the car rolling from stand still:

From Newton's Second Law,

$$F_i = (W/g)^*a = W^*V^2/(2^*s^*g) \, lb_f$$

This is the force (F_i lb_f) necessary to get the car to a velocity (V ft/s) from zero in a distance of (s ft) where $g = 32.2$ ft/s^2.

For $V = 2$ ft/s, $s = 4$ ft, $W = 3,000$ lb, $F_i = 47$ lb to overcome the inertia effect.

Now the force required to overcome the rolling resistance ($\mu = 0.02$) of the tires on concrete is required.

$$F_r = \mu^*W = 60 \, lb$$

So the total push force needed to get the car rolling is:

$$F = 47 + 60 = 107 \, lb$$

This amount of push force is a real effort, but once it gets rolling only 60 lb is required.

However, if you feel strong enough or call a buddy and can get the rolling speed higher, momentum can help again. After all, once the car is rolling fast, you can stop pushing and it will coast to a stop. That's momentum helping you out.

Momentum = m^*V = mass times velocity.

Impulse = F^*t = force acted on for a time t.

Rewriting Newton's Second Law

$$F = m^*a = m^*V/t$$

$$F^*t = m^*V$$

So if a force is applied for a given time, it is equal to the mass times the velocity the mass is moving at.

So for our case,

$$60^*t = (3{,}000/32.2)^*2$$

$$t = 3\,\text{s}$$

$$\text{distance} = V^*t = 2^*3 = 6\,\text{ft}$$

This is how far it will coast due to its momentum. Suppose there was a hill. It would coast up to a stop at different heights depending on the speed when released.

3.8.1 Summary

When you start pushing, you have to overcome the inertia or weight of the stopped car and rolling friction. Once it's rolling, momentum works for you and makes it easier to push.

3.9 Bearing Failure Due To Design Error

A small compact high-performance gearbox was designed for use in a critical application. During load tests on a dynamometer, the small center idler gear bearing kept overheating and failing. The cause was thought to be a lubrication problem, so additional lubrication flow was added which cooled the bearing. The problem was considered to have been solved, until it failed again. Figure 3.9.1 illustrates the system. The idler gears purpose is to change the rotation direction.

The problem was that the designer had based the load on the idler bearing as a load of (P) based on $P = 63{,}000^*\text{HP}/(R^*\text{rpm})$. From a free body diagram, it is shown that the load on the idler bearing is 2^*P. The additional oil may have cooled the bearing, but the heavy load was still present to cause spalling in the small under-sized ball bearings used.

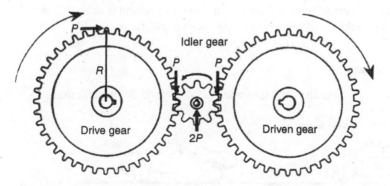

Figure 3.9.1   Load on idler gear bearing.

3.9.1 Summary

Consider all loading possibilities on a new design.

3.10 What Is the Shortest Stopping Distance for My Car?

People seem to drive crazy during holiday seasons. "U" turns in front of you, cars going through red lights or stop signs. It's quite scary. After hearing many screeching tires and seeing cars miss other cars by just 20 ft at 50 mph, I was curious about what was the best stopping distance on dry pavement for an average driver. This doesn't include the driver's reaction time which could be a second or two and could equal the stopping distance. Figure 3.10.1 shows this.

Using an energy balance is one way to determining the stopping distance (d ft). The work energy of the car on the road trying to stop the car is,

$$W_{energyfriction} = \mu^* \text{car weight reaction}^* \text{distance} = -\mu^* W^* d \text{ ft lb}$$

The work energy due to kinetic energy of the moving vehicle that must be dissipated:

$$KE = (1/2)^*(W/g)^* V_0^2 \text{ ft lb}$$

Equating and solving for the stopping distance from some initial velocity:

$$d = V_0^2/(2^*g) \text{ where } g = 32.2 \text{ ft/s}^2$$

For dry pavement and 50 mi/h (73 ft/s) and $\mu = 0.8$.

$$d = 73^2/(2^*0.8^*32.2) = 103 \text{ ft stopping distance}$$

That's a long distance if another car cuts in front of you and is only 20 ft away. Other road conditions will result in longer stopping distances.

3.10.1 Summary

At 50 mph, it takes more than 103 ft to reach a full stop on dry pavement.

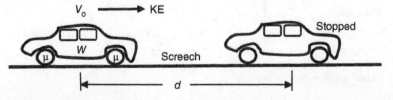

Figure 3.10.1 Stopping a car.

3.11 How Hot Do Brake Disks Get in a Panic Stop?

When making a panic stop, concern was that the brake disks might get so hot that they might warp or distort. A distorted rotor disk would cause a thump, thump, thump when you braked and the rotors would have to be repaired or replaced. In Section 3.10, the shortest stopping distance was determined. With this information, the heat energy put into the automobile rotor disk is known and the rotor temperature can be approximated. Figure 3.11.1 is a typical rotor disk with heat applied by the brake disk pad.

From Section 3.10, the shortest stopping distance for a 5,000-lb automobile from 50 mph was 103 ft. The energy dissipated in friction heat is:

$$W_{\text{friction}} = \mu^* W^* d = 0.8^* 5{,}000^* 103 = 412{,}000 \text{ ft lb}$$

For each rotor disk = 103,000 ft lb or the heat equivalent (778 ft lb/BTU) of $Q_{\text{BTU}} = 132 \text{ BTU}$.

Since the disk is rotating, it is assumed the disk is evenly heated all around. While the rotor weighs 10 lb, only about 3 lb of the rotor disk is being heated.

Recall that the specific heat equation in weight form is:

$Q_{\text{BTU}} = W^* C_{\text{steel}}^* (\Delta T)$ and is the energy required to raise some weight of steel by some temperature differential.

With $Q_{\text{BTU}} = 132 \text{ BTU}$, $W = 3 \text{ lb}_f$, $C_{\text{steel}} = 0.12$, $\Delta T \approx 370\,°F$ or with a starting disk temperature of 100 °F this would be 470 °F.

This is not hot enough to cause distortion of the rotor disk with just one stop. With repeated stops or higher speed stops, the temperature could be much higher.

For example, at 100 mph, it takes 418 ft to stop or about 525 BTU per brake disk. This would result in a disk temperature of 1,500 °F, which is pretty hot but far from the melting temperature of cast iron which is 2,000 °F.

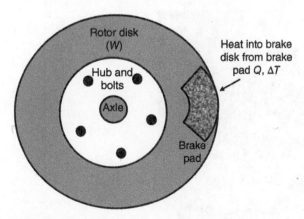

Figure 3.11.1 Brake rotor disk heat input.

3.11.1 Summary

During a panic stop from 50 mph to stop in 103 ft, the brake disk will get to about 470 °F and probably won't warp.

3.12 Will the Turbocharger Disk Go Through its Housing?

We are often near machines that rotate such as flywheels in automobile engines, compressor disks in aircraft, or turbochargers on large engines.

The faster they rotate, the more nervous I am since the kinetic energy of the rotating parts go up with the square of the speed. It's well known that rotating disks can fail and go through their protective housings. Figure 3.12.1 is the bell housing of a racing car whose clutch plate blew up.

I was asked to consult on a turbocharger rotor disk, which had failed on a large diesel engine. It exploded and exited out of the side of the housing and landed 30 ft from the engine. The equipment owner wanted to know if this was a one-time uncontained event or if it could occur again.

Figure 3.12.2 is the analytical model used to help answer the question, along with a lot more research and investigation.

The model is pretty straightforward and general and determines if the force of the broken rotor is enough to punch-out, meaning shear out, the shaded section of the housing shown in Figure 3.12.2.

When the rotor breaks, it will be assumed to depart tangentially.

Figure 3.12.1 Clutch disk explosion through A bell housing.

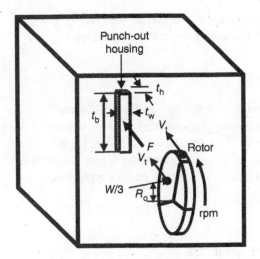

Figure 3.12.2 Analytical model of a failed rotor.

Here the tangential velocity is $(V_t) = 2\pi*R_o*\mathrm{rpm}/60$ in./s and is moved from the outer diameter to the center of the sector.

To punch out a piece of metal, the work involved is:

Work $= F*\delta$ and $F =$ shear stress $(\tau)*$shear area $(A_s)*\delta$

The shear area is the perimeter (L) of the contact zone times t_h.

The perimeter is $L = 2*(t_b + t_w)$.

The kinetic energy of the part in motion is:

$$KE = 1/2*\varphi*W/g*V^2$$

It unlikely that all of the KE will be used in the punching operation, so φ is included as some fraction of the energy. $\varphi = 1$ means all is used and $\varphi = 0$ means none is used.

Equating KE $=$ work done and solving for (δ):

$$\delta = (1/2)* \left[(\varphi*W*V^2/((t_b + t_w)*\tau*g)) \right]^{1/2} \text{ in.}$$

As long as this (δ) is less than the housing thickness (t_h), the part shouldn't punch-out.

This is because the (δ) thickness is all that can be punched out with the KE available.

The shear strength of the housing material is $\tau = 25{,}000$ lb/in.^{2}, rpm $= 10{,}000$, $R_o = 3$ in., $W = 1.4$ lb, $t_h = 0.25$ in., $t_w = 0.125$ in., $t_b = 1.0$ in. After checking various cases, an approximation of $\varphi = 0.5$.

This results in $\delta = 0.40$ in. and with the thickness $t_h = 0.25$ in., this a punch-out condition. Even though the analysis method is very approximate, it does provide useful information. The final solution was to prevent it from failing again by addressing the cause, which was determined to be a rotor material issue.

3.12.1 Summary

Keeping high-speed rotors contained if they fail becomes more difficult as the rotating speeds become higher. Designing to keep stresses low along with periodic inspection for rotor cracks is usually required. For the clutch disk of Figure 3.12.1, $\delta = 0.3$ in. and the steel housing was $t_h = 0.25$ in., which is close to punch-out. From the large hole obviously, much more has occurred than just a simple punch-out. Such failures are complex.

3.13 Failure of an Agitator Gearbox

This failure involved the gears in an agitator that processed rubber slurry in a large tank. A previous failure had occurred where the bolts holding the paddles had bent. The paddle bolt holding system was strengthened. The liquid in the tank was hot, and the product was soft. The product would periodical build up on and between the baffles and tank wall as shown in Figure 3.13.1. After the gear failure, field personnel mentioned that when inspecting this large agitated tank they saw

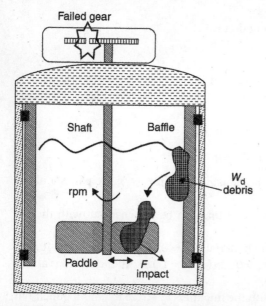

Figure 3.13.1 Agitator tank with baffles.

a chunk of rubber, "as large as a cow hide," as they stated, hanging on the top part of a baffle.

Occasionally when the vessel was opened for cleaning small amounts of debris could be seen between the baffles and tank wall near the liquid level. It was speculated that the sloshing liquid would occasionally knock a large piece loose. As it fell, it would be hit by one of the paddles similar to a baseball being hit by a bat. It would be chopped up into smaller pieces so when the vessel was cleaned, no large pieces were visible. This is one of those phantom failures mentioned in Section 2.10. Here's the analysis used to verify a possible cause.

The paddle tip is moving at some velocity (V), and the falling debris (W_d) velocity is negligible. The question was if this could eventually cause the gearing to fail over time, with enough impacts. The impact torque will be calculated and compared to the mean shaft torque the agitator was designed for.

Consider (W_d) impacting the paddle. The equation for the impact force is after substituting $V = d/t$:

$$F = (W_d/g)^* d/t^2$$

An approximation of the time (t) is the difficult part. The following simplification is made. Here the large piece of debris dropping on the paddle will be considered to be a soft rubber sphere of diameter (d) that sticks to the paddle. As the paddle moves through the sphere (d) at a velocity (V), the time to do this is $t = d/V$ s

$$W_d = (4/3)^* \pi^* (d/2)^{3*} \rho \text{ lb}_f$$

$$\rho = 95 \text{ lb/ft}^3, r = 4 \text{ ft}, \text{HP} = 75, \text{rpm} = 60$$

The impact torque on the mixer shaft and gears:

$$T_{impact} = F^* r$$

$$T_{mean} = 5,252^* \text{HP/rpm}$$

A sensitivity analysis of various size debris is shown in Table 3.13.1.

Table 3.13.1 Debris size and impact.

Debris diameter (ft)	T_{impact}/T_{mean}
1	0.6
2	2.3
3	5.4
4	9.5

3.13.1 Summary

The gears, shafting, and paddles are designed for a maximum $(T_{impact} + T_{mean})/T_{mean} < 3$ which means moderate impacting. It is possible that repeated impacts by falling debris of various size can exceed the design limits of the manufacturer. This would overload the paddle bolts, shafting, and gearing and cause the type failures experienced. Modifications to the baffles which allowed more space between the tank wall and the baffles and shortening the baffles solved the problem by eliminating build-up.

3.14   Failure of an Extruder Screw

A failure occurred on an extruder screw shaft in the high-pressure area of the extruder. The failure was a rotating bending type failure as shown in Figure 3.14.1.

This extruder is used to process a hot product and push it through a die and chop it into pellets. Figure 3.14.2 shows the screw type design. The shaft is solid with the flight spirals machined as part of the shaft. The extruder shaft floats in the barrel on hardened flights and polymer-formed plugs. The hot polymer also acts as a lubricant.

The torque required to move the product produces torsional stress on the extruder shaft. The nominal torsional shear stress S_{ST} due to the torque is a steady-state stress. It is maximum at the drive end and can be approximated with the following equation:

$$S_{ST} = 320,856 {}^* HP / \left[(D_o{}^3)^* RPM \right] \ lb/in.^2$$

Figure 3.14.1   Extruder shaft failure.

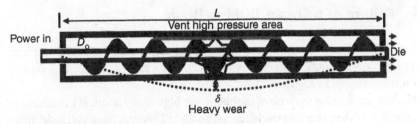

Figure 3.14.2 Extruder model.

Table 3.14.1 Wear deflection and stress.

δ (in.)	S_{ST} (lb/in.^{2})	$S_{b\delta}$ (lb/in.^{2})
0.0325	2,500	±5,000
0.0650	2,500	±9,700
0.0975	2,500	±15,300

HP is the input horsepower, RPM is the shaft speed, and D_o is the outside shaft diameter. Starting-up with product in the screw or "bumping," the drive motor to move stuck material can cause an impact load.

When there is wear in the high-pressure area on the screw flight tips and the barrel, the shaft is pushed down in the clearance (δ) and can increase the bending stress. Every revolution will result in a cyclic bending stress. An approximation of this cyclic stress $S_{b\delta}$ where $E = 29*10^6$ lb/in.^{2} is [1]:

$$S_{b\delta} = \pm 8^* \delta^* D_o{}^* E / L^2 \text{ lb/in.}^2$$

Table 3.14.1 shows the results for $D_o = 6$ in., $L = 96$ in., HP = 150, rpm = 100.

3.14.1 Summary

With a high cyclic bending stress at the center of the extruder shaft due to heavy wear, it is certainly possible that this caused the shaft failure. A periodic wear check preventative maintenance program eliminated any further failures by repairing the screw before it failed. Increasing the wear resistance of the flight hard surface also helped.

Reference

1 Sofronas, A. (2016). *Survival Techniques For The Practicing Engineer*, 141. Wiley.

3.15   Failure of a Steam Turbine Blade

A vibration alarm on a small steam turbine driving a centrifugal compressor occurred. On disassembly, a blade was found missing as in Figure 3.15.1 on the steam turbine shown in Figure 3.15.2.

The failure occurred at the base of the blade, the high stress point, at the interface above the fir tree that attached the blade to the disk. This was the only blade missing and the question was what had caused it? It had happened once before when the turbine overspeed shutdown was being tested. It was suspected this might have occurred again; however, Operations said it hadn't. There was no instrumentation that recorded any overspeed event. This analysis was performed at Operations request. The stress in the failure zone will be examined under adverse conditions along with discussions with the steam turbine manufacturer.

Figure 3.15.1   Missing blade on steam turbine.

Figure 3.15.2 Replacing carbons on 3,000 HP-steam turbine.

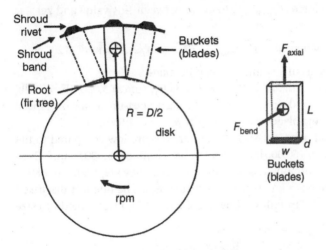

Figure 3.15.3 Analytical model of blade failure.

Figure 3.15.3 represents a sketch of the model used for the analysis. The metallurgical report stated the failure was a fatigue failure, and there was no evidence of any corrosion or defects. The report didn't define the origin but suggested it probably started at a very small defect as the failure was defined as a high cycle fatigue type.

There are two primary cyclic loads on the blade that can cause fatigue. The first is the axial load F_{axial} normally caused by the centrifugal force change during start-up or speed cycling. Overspeed would affect this stress but would only be one cycle. The second is the bending stress, which is due to the bending load F_{bend} as each blade passes a nozzle. The effect of this load is greatly reduced because of the shroud band attached on the blade top. With the shroud, the assembly acts as one large cantilever beam instead of individual small ones. What happens if the shroud band fails and a blade is unrestrained? That is what will be examined.

The axial force on a blade due to centrifugal force and the axial stress are,

$$F_{axial} = 28.4^* W^* (D/2)^* (rpm/1{,}000)^2$$

$$\sigma_{axial} = F_{axial}/(w^* d)$$

This force is trying to stretch the blade.

The bending force on a blade due to gas pressure can be approximated by determining the torque on each disk. It is assumed that the three stages see 60%, 30%, and 10% of the total torque and therefore the force. The torque (i.e. $T = F^* D/2 = 63{,}000^* HP/rpm$), and, therefore the force, is divided by the number of blades per disk (n) for the force per blade.

$$F_{bendblade} = 2^* 0.6^* T/(D^* n)\,\text{lb}$$

This causes the blade force ($F_{bendblade}$) to cycle between this value and zero as the blade passes the stationary nozzles.

$$\sigma_{bend} = k^* M^* c/I, \text{where } M = F_{bendblade}{}^* L/2, c = (d/2), I = w^* d^3/12$$

With some initial data, a cursory analysis can be made.

$L = 2\,\text{in.}$, $d = 0.5\,\text{in.}$, $w = 1\,\text{in.}$, $D = 40\,\text{in.}$, $HP = 3{,}000$, $rpm = 5{,}000$, $W = 0.3\,\text{lb}$, $k = 2$, $n = 150$, $\sigma_{axial} = 8{,}500\,\text{lb/in.}^2$ one cycle each start-up or an overspeed $\sigma_{bend} = 0\text{–}400\,\text{lb/in.}^2$ during blade passing or $rpm^* n/60 = 12{,}500\,\text{cps}$.

Considering no shroud band is assumed, the stresses are low compared to the material tensile strength of $150{,}000\,\text{lb/in.}^2$ and fatigue limit of $75{,}000\,\text{lb/in.}^2$. Even considering a stress concentration factor of two stresses, they are relatively low.

Failures of this type happen on all size steam turbines. The one just discussed was a small 3,000 HP unit. The failure shown in Figure 3.15.4 was a medium size

Figure 3.15.4 Steam turbine 40,000 HP rub failure.

40,000-HP unit that failed due to a rub. Notice this failure also occurred in the root (i.e. fir-tree).

3.15.1 Summary

Normally in these types of failures, debris, corrosion, a rub or a small defect would have been noticed; however, this wasn't the case. The failure was definitely high cycle fatigue and not a one cycle sudden overload as would be expected from overspeed. Operations said an overspeed hadn't occurred and there was no indication one had. One thought is that the shroud band was defective, cracked, and fluttered and the blade vibrated due to eddies and other flow effects not analyzed. After long discussions with the manufacturer, it was learned that their newer turbines have replaced the shroud band rivets with a new design. This suggests the shroud band may have failed first and caused the subsequent blade failure.

The new shroud rivet design was used on the reblading of the spare rotor. With seven years of operation, the installed rotor ran successfully until the steam turbine was replaced with a high-efficiency design.

3.16 How Long Will It Last?

An analytical model can be like a time machine as it lets you look into the past and into the future. This of course doesn't work on many things, for example the stock market, as discussed in Section 10.14.

Just think of the benefit this has to an engineer. A crack is noted and you are asked if the equipment can continue to operate until a planned shutdown can be organized. This may be because an unplanned shutdown will cause lost production or contracts and the wrath of the customer. Saying production should be shut down immediately without any supporting data won't make an employee very popular. This could even be worse if they decide to continue to operate for a month and the crack never grows. The employees' advice may not be requested in the future because of the loss of credibility.

There are good reasons to be cautious. Brittle fracture is a failure that occurs in brittle or defective steels and other materials. An example might be to break a piece of chalk in bending. Unlike the ductile paper clip bending model which can be bent many times, the chalk snaps immediately in one cycle. Steel also can fail in this manner especially in older steels. The crack extends at 5,000 ft/s, so there is no time to measure the growth trend. It may grow slowly at first and then zip through 10 in. of steel in a fraction of a second. The continued use of old steels is analyzed using brittle fracture techniques and comparing the stresses with the materials toughness.

After performing many of these types of evaluations, it became obvious that this growth could be utilized on ductile materials as well. Now it would be a stable growth due to the ductile nature of the material and the limit would be when the material's strength could no longer support the load and it failed.

The following example describes this technique to help in making decisions when the material is a new type steel known not to be brittle [1].

After a sudden shutdown and restart of a large synchronous motor, the gearbox was inspected as a precaution. A crack was noticed on the edge of one of the teeth similar to the one shown in Figure 3.16.1 but much smaller. It was not known when it had occurred, but it was suspected it may have started because of the power outage and start-up logic. The following analysis was performed before stating and opinion.

If the tooth breaks off and falls into the gear mesh, a catastrophic failure, meaning cracked gearbox, failed bearings, broken shafts, gears, and couplings, could occur. Some have said the tooth will be thrown clear of the mesh, others say it won't. It's impossible to know. The safe call is to replace the gear; however, this was a complicated gearbox and required a several day production outage to disassemble, install, and reassemble the gear shafts. Management wanted to know if they could finish the production run with the crack, which would take three more days.

Traditional fatigue design and endurance data won't be of use here since in these approaches almost all the fatigue life is used for crack initiation. Here the tooth already had a crack in it. Figure 3.16.1, which is not the actual tooth but the depth of the crack, may have been in the order of 1/32 in. to be conservative.

Figure 3.16.2 shows a single tooth with an initial crack (a_o) across the face (b) on the pinion gear root under the operating load (F):

The simplified load (F) at the pitch diameter (D_p) with (n) teeth engaged trying to bend the cracked tooth is:

$$F = 126,000\, \text{HP}/(D_p * \text{RPM} * n)\, \text{lb}$$

Figure 3.16.1 Typical tooth crack at root.

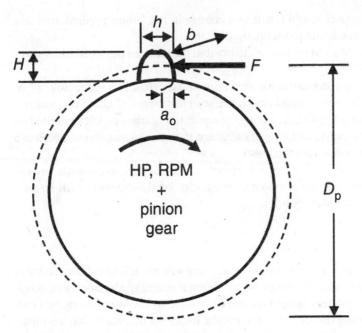

Figure 3.16.2 Pinion tooth with load and crack.

The nominal bending stress at the root of the tooth at the crack, which is a cyclic stress from zero to σ_b, trying to open the crack once per revolution is:

$$\Delta\sigma_b = 6^*F^*H/(1{,}000^*b^*h^2) \text{ ksi}$$

For many ductile steels, the number of cycles for a surface crack to grow from (a_o) to (a_f) with a cyclic stress of $\Delta\sigma_b$ is approximated as [1]:

$$N = [\{8.3^*10^8\}/\Delta\sigma_b^{3.25}]^* [1/\{a_o^{0.625}\} - 1/\{a_f^{0.625}\}] \text{ cycles}$$

For this case, horsepower is HP = 4,000, RPM = 1,500, h = 1.0 in., D_p = 39 in., H = 0.9 in., face length b = 6 in., n = 1.5.

The initial crack length was measured as a_o = 0.033 in., and it is assumed that the tooth will break off when the crack is halfway through or $a_f = h/2$.

Inserting these values into the three equations yields:

$$\Delta\sigma_b = \pm 7.0 \text{ ksi}$$

This value is significant because the threshold for crack growth is approximately ±3 ksi, so this crack will grow.

$$N = 12.2^*10^6 \text{ cycles or } N/(\text{RPM}^*60^*24) = 6 \text{ days}$$

With the crack growing halfway through the tooth in six days, there is a high risk of failure in continuing to operate the gearbox. Slight changes in the crack size or tooth load assumptions could result in a change in the failure prediction to one day of operation before the tooth breaks off.

The method is not accurate enough to say that the unit can be operated for three days and not fail.

This analysis was necessary since a prominent individual was approving operation by stating, "It's only a small crack, I've seen this before!." These calculations made taking a 3-day production outage to repair the gearbox much easier to justify than trying to explain a two weeks outage to senior management because a cracked tooth had fallen into the mesh.

When someone has performed a reasonable analysis their arguments usually carry more weight than those who are speculating on the outcome with no supporting data and only their opinion.

3.16.1 Summary

In this example, the gear with the crack was sent to a metallurgical laboratory to determine how the gear failed and when. Using a scanning electron microscope (SEM), the gear tooth was broken off and examined. By counting the fatigue striations and analyzing the surfaces, the results indicated that there was an initial impact at about the time of the power surge which started the initial crack. It then progressed by the cycling fatigue load. It was predicted that it would continue to grow at a faster rate as the stress increased due to the reduction in cross-sectional area as the crack grew in depth. The report stated it would have failed soon.

This type analysis was used to make decisions on several other cases which were:

- crack growth in a rotating shaft
- paddle failures on a feeder
- crack growth failure on a large disk dryer
- pitting across the face near the root on a ships gear
- small crack ground out of an aircraft crankshaft

Reference

1 Sofronas, A. (2006). *Analytical Troubleshooting of Process Machinery and Pressure Vessels*, 339. Wiley.

3.17 Gear Life With a Load

Destructive pitting across a pinion face as in Figure 3.17.1 or broken gear teeth are the type failures the author has analyzed many times. The failure can tell you

Figure 3.17.1 Destructive pitting.

much about what happened, especially when it's combined with a mechanical and metallurgical analysis.

Gears are designed to fail in pitting rather than tooth breakage in bending. This makes sense as breakage can result in a sudden catastrophic failure. Pitting occurs over time and noisy gears, vibration or metal in the oil are usually detected before a catastrophic failure occurs. Pinions usually see pitting first because they rotate faster and experience more stress cycles.

The following analysis has been used to help determine possible causes of pitting and as a means of documenting historical failures.

The contact stress on a pinion face of a typical industrial quality gear is:

$$S_c = (2.76^*10^6/D_p)^* \{[(M_g + 1)/M_g]^*[HP/(RPM^*F)]\}^{1/2} \text{ lb/in.}^2$$

M_g = number bull teeth/number pinion teeth.

The failure life can be related to the contact stress on the pinion face using Figure 3.17.2. The data points are where the author has experienced destructive pitting using the abovementioned equation. Most pinions were through hardened spur or helical but included some case hardened herringbone.

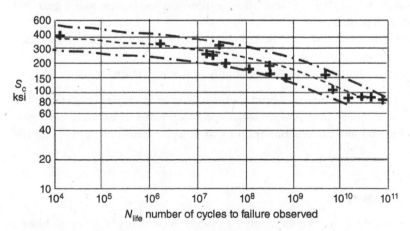

Figure 3.17.2 Contact stress and cycles to failure.

These points are actual failures and are cause for concern and would justify a thorough analysis using gear design programs. Points above 10^{10} were associated with normal life wear-out while contact stresses above 300,000 lb/in.^{2} resulted in plastic deformation of the pinion face or tooth breakage.

Consider the case where destructive pinion pitting was noticed on a new ship gearbox after only 8,000 hours of operation. The pinion has a face width $F = 8$ in., horsepower HP = 2,000, RPM of the pinion is 1,000 and has 40 teeth, the bull gear has 200 teeth, and the pinion diameter D_p = 11 in.

$$S_c = 137,000 \text{ lb/in.}^2$$

From Figure 3.17.2 while there is a wide scatter band, consider a point $N_{life}/(60*RPM)$ hours $\approx 8*10^9/(60*1,000) = 133,333$ hours.

This is an acceptable life for this application. Why was it pitting at 8,000 hours?

The equation can be helpful as HP, RPM, F, and D_p are included. Poor face contact due to misalignment or load sharing in the case of herringbone gears can easily reduce the effective face width F which increases the contact stress and reduces the life.

In this example, load sharing was the problem, and only half of the face was showing contact. There was no pitting on the unloaded back face, so torsional vibration was not the cause of the pitting. This resulted in a contact stress of 194,000 lb/in.^{2} or a lower bound life of only $5*10^8$ cycles (8,330 hours). Correcting the load sharing solved the problem.

When a gear fails, an investigator typically,

- performs a full gear evaluation to determine if the basic design was adequate especially after a re-rate.
- performs a metallurgical examination on the hardness profile of the gear and material properties.
- examines the wear pattern to determine if there was load, misalignment, bearing, or contamination issues.
- examines the lubricant type, filters, supply, and temperature for adequacy.

Gear design is a complex subject with loads, gear geometry, materials, and lubrication all interacting. Always seek expert help when assessing critical gear failures but do some calculations yourself, so you can talk intelligently with the experts.

3.17.1 Summary

Having quick methods to help identify a failure cause is useful to avoid having speculation control the investigation. While the analysis method may not be totally correct, it can add new data to a failure investigation.

3.18   Analyzing the Life of a Gear

In Section 3.17, it is shown that a simple analysis can be performed to see if a high contact stress was the cause of pitting on a gear face. It was used to help determine the cause of the pitting. For example was the load too high or the gear face contact in error. This is a quick failure test based on the engineers' experiences.

When a new gear system is being reviewed for adequacy, this type analysis is not appropriate. Sophisticated analysis programs are used to evaluate the many types of failure mechanisms possible and to see if the gear system is adequate. These are usually Code based and therefore experienced based also. This type of analysis may be performed by a consultant when a company wants an independent analysis to verify the acceptability of a new gearbox.

An analysis must be done on each gear set. The gear types can be spur, helix, herringbone, spiral, or most other designs. Planetary gear systems may be quite complex, but after determining the gear speeds and loads, they are treated in a similar manner. Gear material properties and the lubrication films developed with specific lubricants are all part of the analysis.

Criteria based on experience are used to examine the possibility of bending fatigue, pitting fatigue, scoring, and wear. By using a Code-based program, such as one based on the AGMA (American Gear Manufactures Association) rating analysis, the experience and analysis of contributors can be utilized in the programs. Most countries have their own specifications and programs, which are all usually quite similar. After all each does not want the gears to fail and the mechanism is similar.

Consider the following example on one pinion in a large complex gearbox with input horsepower 10,000 and a pinion rotating at 360 rpm. The pinion has a pitch diameter of 23.6 in. and a face width of 18.9 in. The gear has 37 teeth and pinion 37.

An analysis considering all of the pinion loads, dimensions, pinion material, and lubrication properties might produce summarized data such as shown in Table 3.18.1.

This indicates the pinion has a relatively short service life. A longer life would be possible at lower loadings or with gear tip relief modifications. It is not optimum and is why these types of calculation reviews should be done before the gearbox is manufactured. Only in this way can improvements be suggested.

Table 3.18.1   Partial AGMA type calculations.

Specific tooth load (lb/in.)	Allow/actual bending stress	Allow/actual contact stress	Bending life years	Pitting life years	Probability of wear	Probability of scoring
8,000	1.5	1.3	OK	4.8	9%	<5%

The AGMA calculations contain factors that improve the resistance to pitting and would produce a longer acceptable life. The specific tooth loading also seems high. A similar analysis would need to be performed on each gear in this gearbox, including the planetary gearing.

A check using the calculation method of Section 3.17 would show a contact stress of 202,000 lb/in.^{2} and a life of $N_{life} = 5*10^8$ cycles or $N_{life}/60*rpm = 23,000$ hours or 2.6 years. This is not based on a newly designed, high-quality gear and is why the life is relatively low.

3.18.1 Summary

Independent gear calculation reviews of a new gearbox should be performed at an early stage before the gearbox is built. Only in this way can improvements to the design be made. Many times the Owner has other gearboxes that have included modifications that have proven successful. Slight gear modification can sometimes greatly increase the reliability of the gears box. These should be included in the purchase specifications for new gearboxes.

3.19 Predicting the Cause of a Gear Tooth Crack Growth Past and Future

In engineering, we sometimes wish we could see what is going to happen in the future or why something has happened in the past. This may occur when management questions why something has occurred and what the risk is in continuing to operate until a planned downtime. They want you to predict the past and the future.

Consider the following case history. A crack is noticed on a gear tooth after a sudden power outage.

The risk is that if the tooth breaks off and falls into the gear mesh a major wreck could occur resulting in a long outage. Knowing when and why the crack started is also important, so a repeat event won't occur. A start-up after gear replacement without knowing the cause of the crack is always risky.

Crack growth type equations can be used on ductile steels. Knowing or assuming an initial crack size and knowing the cyclic stress trying to open the crack, the cycles to grow the crack to a certain length can be estimated.

For this case, let's assume some event started a small crack in the tooth root across the tooth face as shown in Figure 3.19.1. It takes a very large shock load of many times the allowable stress to cause a crack to develop through a case hardened gear tooth [1]. The case depth may be $(a_i) = 0.05$ in. deep in the tooth face width (b). When the crack was first noticed during an inspection, it appeared to

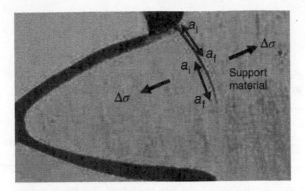

Figure 3.19.1 Crack in gear tooth.

have grown to a 0.13 in. length. When this crack propagates half-way through the 1 in. wide tooth $(a_f) = 0.5$ in. as shown, it will break off in a bending type failure. This is because as the crack grows it reduces the support material.

The cycles (N) to grow the crack can be approximated by integrating the Paris Differential Equation $da/dN = C^*(\Delta K)^m$. Here (da) represents the crack growth during the cycle (dN) and C and m are material constants [2].

$$N = \left(8.3^*10^8/\Delta\sigma^{3.25}\right)^* \left(1/a_i^{0.625} - 1/a_f^{0.625}\right) \text{ cycle}$$

This equation is for a large plate where the cross-section doesn't significantly change as the crack grows. The constant surface bending stress $\Delta\sigma$ from 0 to peak stress can be used.

This is different for a rectangular beam tooth as shown in Figure 3.19.2. The stress becomes an iterative procedure at each growth step since the bending resistance of the supporting material becomes less as the crack grows. Only two growth steps are used in this example with the stress average used for each step shown in Figure 3.19.2. As is shown in Table 3.19.1 just using the surface nominal stress would not change the recommendation to repair before start-up.

Table 3.19.1 shows the results from these calculations.

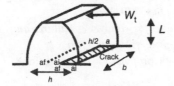

$\Delta\sigma_{surface} = 6^*Wt^*L / (b^*h^2)/1{,}000$ ksi

$\Delta\sigma_{crack} = 6^*Wt^*L / (b^*h{-}a)^2)/1{,}000$ ksi

Figure 3.19.2 Gear tooth crack growth.

Table 3.19.1 Crack growth with stress.

Cyclic stress on tooth	a_i (in.)	a_f (in.)	N	Life at 1,200 rpm
$\Delta\sigma = 11.6$ ksi to start $a = 0.09$ in. average	0.05	0.13	8.7*105 cycles	12 h ago
$\Delta\sigma = 16.9$ ksi to fail $a = 0.31$ in. average	0.13	0.50	1.8*105 cycles	2.5 h to fail
$\Delta\sigma = 10$ ksi start to failure	0.05	0.50	1.3*106 cycles	18 h to fail

From this data it appears that the power outage [3] could have been the cause since it can result in a large gear tooth shock load. Once started the crack grows very quickly so caution suggests not starting the unit up until the gears are replaced. Thus, you have looked into the past and also into the future.

This is certainly a very approximate method, but for the person involved, it does lower the risk in making a decision. Cracks can grow very quickly with a sizable cyclic stress as testing has shown [4] and can be in the $1*10^5$ cycle range to break-off a gear tooth.

When someone wants to override your recommendation, at least you have documented how yours was made and the responsibility is now theirs. When they decide to run for a week and the gear survives, their decision was based on a guess and yours was based on using engineering principles.

The important points of this chapter are that a crack in case hardened gear teeth can be expected to grow rapidly and fail after a hard impact, in a matter of hours not days. When someone says, "We can run a week, I've seen cracks like this before, let's start back up!," you shouldn't agree without your own data to substantiate that claim.

References

1 Handschuh, R. et al. Investigation of low-cycle bending fatigue of AISI 9310 steel spur gears, NASA/TM-2007-214914.

2 Sofronas, A. (2006). *Analytical Troubleshooting Of Processing Machinery And Pressure Vessels Including Real-World Case Studies*, 339. Wiley.

3 Sofronas, A. (2021). *Unique Engineering Methods for Analyzing Failures and Catastrophic Events: A Practical Guide for Engineers*. Wiley, TBP.

4 Lewicki, D. and Ballarini, R. (1996). Gear crack propagation investigations. *Technical Rep. ARL-TR-957*.

3.20   Nonlinear and Linear Impact Problems

Many times we have to solve engineering problems from available data. Possibly it's for a new design or to analyze a failure. Here is a method that has been found valuable to obtain the force of a nonlinear impact. While more complex tools are available, this method can produce adequate results when the data is available. It has been used successfully on impact barriers, polymer packing in extruders, crushing of plastics and metals, and other areas in the processing industry.

Nonlinear impact problems can be solved if the nonlinearity can be expressed graphically with static load and deflection data. Consider the crushing of an automobile. It may have a curve such as shown in Figure 3.20.1. Notice how it differs from a linear curve and the load gets very large for the nonlinear case as the deformed product approaches a solid block. The slope at any point on the curve, which is also the spring constant, changes. The linear case is a constant. This nonlinear case might be from a spring bottoming out or material packing up solid like the crushing of an automobile impacting a barrier wall as shown in Figure 3.20.1.

P_{static} represents the resulting static load from a δ_{static} static displacement. The data is usually available in tabular form with the load taken at each displacement step.

For the linear case, the spring constant is $k = P/\delta$.

The load in the spring is therefore $P = k*\delta$.

The energy under the linear curve is the area or:

$$E_{linear} = 1/2 * P * \delta$$

For a nonlinear curve, it's more complex and the slope changes along the curve and represents the spring constant at different points on the curve.

When the equation of the curve is known calculus can be used to determine the slope at any point which is also shown in Figure 3.20.1.

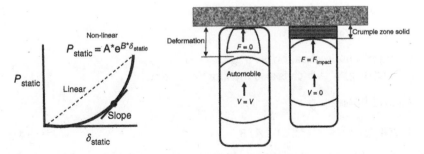

Figure 3.20.1   Linear and nonlinear load displacement.

When the data can be fit to an equation of a curve in an exponential form to the base (e) with (A) and (B) constants, it results in (Eq. (1)):

$$P_{static} = A^* e^{B^* \delta static} \tag{1}$$

The tangent to this curve at any point is the derivative and is the spring rate at that point:

$$k = dP/d\delta = A^* B^* e^{B^* \delta} \tag{2}$$

Inserting (δ) will produce the spring constant at that point and the force is:

$$F_{spring} = k^* \delta = A^* B^* \delta^* e^{B^* \delta} \tag{3}$$

The energy under the nonlinear curve is,

$$E_{nonlinear} = \int A^* e^{B^* \delta} = A^* e^{B^* \delta}/B - A/B \text{ from } \delta = 0 \text{ to } \delta = \delta$$

All that is needed is to fit the data to an equation using a curve fitting technique such as is available in software such as Excel. The exponential curve fit will produce the constants (A) and (B) and also R^2. You can think of R^2 as a goodness of fit, and has a value of 0–1. Zero means the fit is no better than a horizontal line going through the mean of the data which is not of much use. A value of 1.0 indicates a line going through each of the points. A good value for R^2 is in the 0.95 range. When it is less, more data is usually required or erroneous points removed for a good fit or a form other than $P_{static} = A^* e^{B^* \delta}{}_{static}$ needs to be fitted using another method than presented here.

Next to be determined is the force developed during deceleration of an impacter moving at velocity (V) of weight (W) to a depth (δ):

$$F = m^* a = (W/g)^* a$$

But the deceleration is the time it takes the velocity to go from (V) = (V) to (V) = 0 over the distance (δ):

$$a = V^2/(2^* \delta)$$

$$F_{decel} = (W/g)^* V^2/(2^* \delta)$$

The energy of this deceleration is $F_{decel}^* \delta$:
$E_{decel} = 1/2(W/g)^* V^2$ which is also the kinetic energy.

$$E_{decel} = E_{nonlinear}$$

$$1/2(W/g)^* V^2 = A^* e^{B^* \delta}/B - A/B \tag{4}$$

Equation (4) can be solved for the crush depth (δ) by iterative means and putting (δ) back into (Eq. (3)) solves for the impact force:

$$F_{impact} = k^* \delta = A^* B^* \delta^* e^{B^* \delta} \tag{5}$$

Table 3.20.1 Automobile crush test data.

P(lb)	10,000	20,000	25,000	55,000	80,000	125,000	200,000	400,000	550,000
δ(ft)	0.50	1.00	1.25	1.50	2.00	2.50	3.00	3.75	4.00

The time for the impact to reach the displacement (δ) is:

$$t = \delta/V \text{ s}$$

Here's an example of its use when an automobile crashes into a barrier wall.

Assume the data in Table 3.20.1 is obtained from a static crush test, meaning a hydraulic press deforms the metal at a slow speed.

From curve fitting $A = 7221.4$ and $B = 1.1047$ which results in a $R^2 = 0.98$.

The question is what is the impact force when this 3,000-lb automobile impacts the barrier at 51.3 ft/s (35 mph).

With $W = 3,000$ lb, $V = 51.3$ ft/s (35 mph), and $g = 32.2$ ft/s^2:

$$F_{impact} = 418,000 \text{ lb}, \delta = 2.68 \text{ ft}, t = 0.053 \text{ s}$$

While only an approximate method, it is quite versatile and easily applied to many different cases.

Now for comparison consider the spring to be linear as shown as the dashed line in Figure 3.20.1, which is far from actual, with $k = 550,000/4.0$ lb/ft:

$$F_{impact} = V*(W*k/g)^{1/2} \approx 183,000 \text{ lb which is lower than the nonlinear case.}$$

For this case, assuming a linear deformation would have resulted in a non-conservative value. This is probably because there is less energy absorption under the curve available to dissipate the impact energy. Figure 3.20.2 illustrates

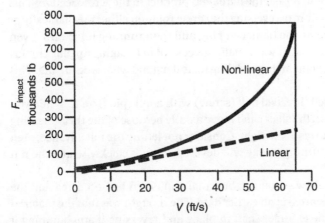

Figure 3.20.2 Impact force on automobile at different speeds.

this as a function of the automobile's velocity. Notice how the impact force gets very large when the material becomes totally crushed. This doesn't occur when a linear analysis is used. Of course the force can't go to infinity as shown as there are factors that limit this such as spring back, barrier movement, and other energy losses such as heat and vibration not accounted for.

3.20.1 Summary

Nonlinear analysis of impact will produce higher impact forces than a simple linear analysis. This is because less energy is dissipated by the impact. A linear analysis is usually adequate for low impact speeds.

3.21 Phantom Failure of an Expander–Dryer

Expander–dryer systems are extruders that remove water from various products by forcing them along an auger-type screw. They exit through die holes at the discharge end of the screw under high temperatures and pressures.

This discusses the failure of such an extruder screw shaft on a 1,000-HP extruder system. The cause of most extruder failures are usually processing issues, extruder bending stresses due to wear, bad bearings, or broken components, typically breaker bolts, wedging in the screw section. In this instance, the failure cause was none of these, and the extruder was repaired and restarted. This unidentified cause was concerning and considered a phantom failure as discussed in Section 2.10.

When nothing was evident on what caused the fatigue failure, it was attributed to "bumping" the motor to unseat the hardened product in the screw section. This is the unapproved procedure of reversing the screw rotation either mechanically or electrically and backing up the hardened plug and then running it forward. Even though it isn't recommended, it was usually successful in breaking up plugs in this type extruder. It is important in the modeling to understand what is occurring. This is shown in Figure 3.21.1.

Consider a left-handed threaded bolt (screw) with a nut (plug) on it. When you turn it counterclockwise, the shaft can't move axially because of the thrust bearing and so the nut (plug) moves to the right. Consider not letting the nut advance, then a large axial force and turning torque will develop. Turn it clockwise and the nut will back up.

The only data available was that an abnormal shutdown had occurred and the extruder system was restarted at an earlier date. The thought was that the material hardened and locked the extruder shaft in place and reversing it and bumping it forward loosened it.

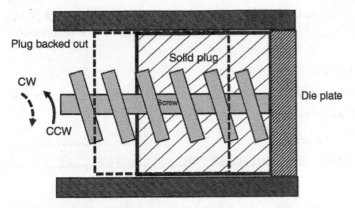

Figure 3.21.1 Moving the plug.

The following analysis determines what impact torques were possible under these extreme conditions.

The motor can attain some rotation because the plug has been backed out and can now be driven forward before it comes to an abrupt stop. With the forward motion, the motor inertia (J) will accelerate until it comes to a sudden stop when the plug is driven back and impacts the die. This kinetic energy will dissipate by winding up the system. In the analysis, the screw is considered to be locked in place as shown in Figure 3.21.2.

Consider that the torque-twist of the screw can be represented by Figure 3.21.2. Since the properties of the solid material are unknown, it will be considered to be holding the screw fixed on the discharge end when it impacts. All the twist is considered to be in the extruder screw, thus the extreme nature of this analysis.

At this point, we can determine the kinetic energy (KE) of the motor rotor due to the "bump" and assume it is absorbed as work done on the extruder shaft (screw).

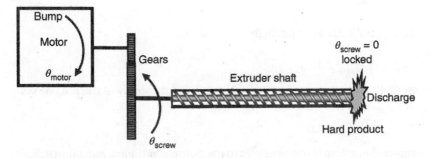

Figure 3.21.2 Locked extruder screw.

The motor is accelerated through (θ_{motor}) from (θ_{motor}) = 0 to (θ_{motor}) in a given time (t) where $\omega = (\theta_{motor}/t)$ and the rotational kinetic energy (KE) available from this is:

$$KE_{motor} = 1/2^*J^*\omega^2 = 1/2^*J^*(\theta_{motor}/t)^2$$

All of this kinetic energy is converted into work done (WD) as it twists the extruder screw through θ_{shaft}.

$$WD_{shaft} = 1/2^*C_{shaft}{}^*(\theta_{shaft})^2$$

These two energies can be equated to each other and solved for (θ_{shaft}):

$$KE_{motor} = WD_{shaft}$$

This (θ_{shaft}) is the value that will make these energies balance with a given (t) and (θ_{motor}) value.

$$\theta_{shaft} = (\theta_{motor}/t)^*[J/C_{shaft}]^{1/2}$$

where $C_{shaft} = 1.18^*10^6{}^*d_{in.}{}^4/L_{in.}$ in. lb/rad where $d_{in.}$ = extruder shaft diameter and $L_{in.}$ is the shaft's length.

This (θ_{shaft}) can then be inserted in $T_{impact} = C_{shaft}{}^*\theta_{shaft}$ to determine the dynamic impact torque due to this "bump."

$$T_{impact} = (\theta_{motor}/r^*t)^*(J^*C_{shaft})^{1/2}$$

A difficulty is in determining what values to use for (θ_{motor}) and (t).

An approximation is available from Operators who historically witnessed the "bumping" procedure. It was usually successful, saved time, and didn't cause any known problems.

The method they followed was about one revolution to back out (i.e. $\theta_{motor} = 2\pi$ rad) and bump time to move forward about $t = 0.2$ seconds.

What impact torque could have been developed from reversing and bumping forward when compared to the full load torque?

$$J = 2{,}000 \text{ in. lb s}^2, \text{ HP} = 1{,}000, \text{ rpm} = 200, \ d = 12 \text{ in.}, \ L = 200 \text{ in.},$$

$$r = 5, \text{ gear ratio.}$$

$$T_{fullload} = 63{,}000^*HP/rpm \text{ in.lb}$$

$$T_{impact}/T_{fullload} \approx 10.0$$

Assuming that for some reason, possibly torsional flexibility of the plug or system, that C_{shaft} is 10 times smaller:

$$T_{impact}/T_{fullload} \approx 3.0$$

This shows the system is very sensitive to the bump conditions, meaning (θ_{motor}) and (t) along with the system rigidity (C_{shaft}).

3.21.1 Summary

This very approximate analysis provided some guidance when little data was available. It shows that with the assumptions made, bumping of a motor to break loose hardened product could result in damaging impact loads, meaning it might start cracks in shafts, gear teeth, or couplings. It could also develop a large axial force on thrust bearings. Continued operation under normal full torque could cause this crack to eventually grow into a fatigue failure.

Even though no validation data was available to confirm this, the cold start-up procedure was modified. This ensures that the extruder barrel is clear of hardened product before a restart is attempted. A precaution against bumping was included in the start-up procedure and also put on a placard on the motor. While not conclusive that bumping was the cause, no unexplained failures have occurred on this extruder for 10 years using the new cold start-up procedure.

Bumping of an extruder to loosen hardened material can result in high axial loads and torques on an extruder screw and should be avoided.

3.22 Cracking of a Rail Hopper Car Due to Couple-Up

The petrochemical industry has all types of machinery, vessels, structures, and transportation equipment. Unfortunately, they can all have failures which can limit production or delivery to customers. The cause of these failures has to be determined and rectified, so that they don't reoccur.

As mechanical engineers, we are sometimes asked for our opinions on a subject which we may not be that familiar with. Fortunately, research and analysis can provide additional data to make an informed recommendation.

In this case, hopper-type railcars that were owned by a chemical company and carrying polymer pellets were periodically developing cracks in their structures. The Yard Superintendent, whom I knew, asked for an opinion on what could be causing the cracking. The only way to make an educated comment on a question like this was to gather some data and perform a simple analysis on the actual system.

Reviewing the literature revealed that high-speed couple-up damage to railcars was known to have occurred based on experimental tests [1]. A yard technician who observed this told the Superintendent that it appeared to be quite violent and much faster than normal. Figure 3.22.1 shows the coupling of two rail cars.

The A.A.R. (Association of American Railroads) recommends a maximum couple-up speed of less than 4 mi/h. Damage from field tests indicates structural damage to railcars at a couple-up velocity of over 9 mi/h [1].

These couplers have an internal compression member such as springs or compressed rubber which compresses to absorb the shock of impact. Too fast and it

Figure 3.22.1   Two rail cars coupling.

is theorized that the spring in the coupling could "bottom-out" or go solid. This would cause extreme loads. Think of hitting a steel table with a hammer and then putting a thick piece of soft rubber on the table and hitting that. No rubber would be similar to the "bottom-out case." The analytical model developed will be used to evaluate this possibility.

This spring-mass system shown in Figure 3.22.2 represents two masses impacting each other through a flexible spring element. Figure 3.22.2 illustrates the collision with velocities before it occurs (u_1, u_2) and after (v). What is needed is the force (F) in the spring (k_{equiv}) and deflection (Δx) after the collision.

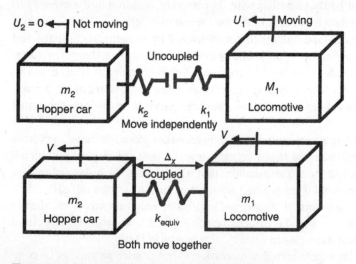

Figure 3.22.2   Two mass collision before and after.

For an inelastic collision for the case $u_2 = 0$ and after solving the conservation of momentum equation considering they move as one after the collision, the final velocity (v) of the two masses moving together is,

$$m_1{}^*u_1 + m_2{}^*u_2 = m_1{}^*v + m_2{}^*v$$

$$v = (m_1/(m_1 + m_2))^*u_1 \text{ ft/s}$$

Equating the final system kinetic energy to the work done on the spring element (k_{equiv}):

$$1/2{}^*(m_1 + m_2)^*v^2 = 1/2{}^*k_{equiv}{}^*\Delta x^2$$

$$\Delta x = v^*((m_1 + m_2)/k_{equiv})^{1/2} \text{ ft}$$

The force in the spring and on each mass is by Newton's Third Law:

$$F = k_{equiv}{}^*\Delta x \text{ lb}_f$$

$$\Delta x_1 = F/k_1 \text{ ft}$$

$$\Delta x_2 = F/k_2 \text{ ft}$$

The equivalent spring (k_{equiv}) is just combining springs in series:

$$k_{equiv} = k_1{}^*k_2/(k_1 + k_2) \text{ lb}_f/\text{ft}$$

This can be applied to the two rail cars coupling up and the coupling force and spring deflection at different coupling speeds using the following data:

$$k_1 = k_2 = 1.8^*10^6 \text{ lb}_f/\text{ft}, m_1 = 9,320 \text{ lb}_f \text{ s}^2/\text{ft}, m_2 = 6,200 \text{ lb}_f \text{ s}^2/\text{ft}$$

For reference $W_1 = m_1{}^*32.2 \text{ lb}_f, W_2 = m_2{}^*32.2 \text{ lb}_f$

Figure 3.22.3 illustrates the forces and deflections at various couple-up speeds based on the equations.

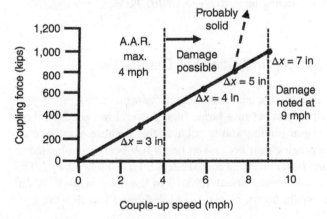

Figure 3.22.3 Analysis couple up speed versus force and spring compression.

Figure 3.22.3 indicates that it is possible to have severe structural distress of a railcar with speeds higher than 9 mi/h. Discussing this with the technician, she said they coupled-up at a speed which she could run at or about 8 mi/h but normally couple-up at about a walking pace. This would be 2–4 mi/h. This seems to confirm her comment that it was a violent couple-up. With a heavy impact, there is a large displacement of the spring (Δx) with the possibility of it becoming solid. This could result in extreme impact loads as shown in Figure 3.22.3.

The recommendation made to the Yard Superintendent was to provide new yard and car signage to limit couple-up to 4 mi/h maximum and for additional training, as too high a couple-up speed could cause the cracking observed.

Consider all that was obtained from this simple analytical model and reference [1].

- The impact load.
- The speed at which damage could be expected.
- The coupling deflection and "bottoming-out."
- The acceptable couple-up speed.
- The effect of the railcar loaded and empty. (i.e. little effect and doesn't change recommendation)
- The ability to make a logical recommendation based on good data.

That was much-needed data to help resolve the issue.

Reference

1 Llana, P., Jacobson, K., Stringfellow, R. (2019). Conventional and crash energy management locomotive coupling tests. *DOT/FRA/ORD-19/36*.

3.23 Loss of Oil Supply and Gear Set Destruction

I recently witnessed the most severe gear destruction I've seen during my 55-year career. It was on the output gear set in a large, high power, low-speed gearbox. Unfortunately, the actual gear photograph is not available because of client confidentiality. The original pristine gear set looked like it had been forge-heated to a cherry red condition and beaten with a hammer as shown in Figure 3.23.1. The temperature under the hammer was about 1,800 °F in the color version. When cooled, the surface was rough, burnt, blacken, and deformed just like the gear set was.

The temperature was so hot that all of the teeth had become malleable, distorted, and had yielded.

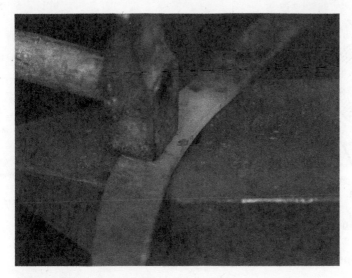

Figure 3.23.1 Forging temperature steel bar.

The question asked by the client was, "Would operation without lubrication for 30 minutes cause this type of damage? Why were the adjacent supporting roller bearings still able to spin?" Answers would be helpful in determining the cause and rectifying it and the following simplified analytical model used to help in doing this.

Consider that the load-sharing gears are producing sliding friction (F_f) on the gear teeth because of the operating load (F_n), $F_f = \mu * F_n$ as shown in Figures 3.23.2 and 3.23.3. All gears do this and with a good hydrodynamic oil film this sliding is usually minimal. A rolling action over the gear profile produces most of the driving force. The friction coefficient (μ) is based on the hydrodynamic film between the teeth. A poor profile, not enough lubrication, incorrect or too thin of

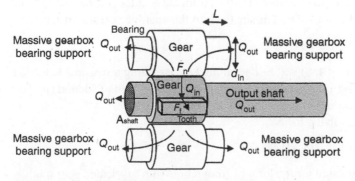

Figure 3.23.2 Heat into gear model.

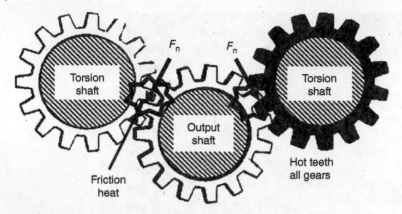

Figure 3.23.3 Heat going into gears.

a lubricant would normally cause a higher temperature along with scuffing and excessive wear.

Scuffing operating without lubrication is not considered in the design of a gear because it's such an abnormal event. Low pressure/low flow shutdowns or temperature monitoring usually prevent this.

The temperature of gear teeth without lubrication and the time to reach this temperature are what will be determined from the following analysis. A gear set mesh of this size usually requires at least 8 gal/min of oil to prevent scuffing [2].

The operating force on a gear tooth is [3]:

$$F_n = 126{,}000{*}\mathrm{HP}/(d_{in}{*}\mathrm{rpm})$$

This can be converted into the heat developed on a tooth:

$$Q_{in} = 4.6{*}\mu{*}F_n{*}v \ \mathrm{BTU/h} \tag{1}$$

The value (v) is the sliding velocity of the gear profile and is 1 ft/s for this gear set.

The heat generated will be assumed to flow from the gear teeth through the shaft by conduction and absorbed and dissipated into the massive gearbox structure.

$$Q_{out} = k{*}A/L{*}\Delta T \ \mathrm{BTU/h}$$

The heat flow path (A/L) will be determined by assuming a reasonable loaded gear temperature with a typical gear hydrodynamic coefficient of friction (μ).

Using a power balance with six paths to dissipate Q_{out}:
$Q_{in} = 6{*}Q_{out}$ and solving for ΔT:

$$\Delta T = 4.6{*}\mu{*}F_n{*}v/(6{*}k{*}(A/L))°\mathrm{F}$$

A kinetic sliding value for (μ) for a hydrodynamic film lubricated gear tooth is [1] $\mu \approx 0.02$ and for a non-lubrication surface can be $\mu = 0.1$ or higher.

The temperature rise (ΔT) based on the friction coefficient is what will be analyzed.

The initial data for this system is as follows:

HP = 2,750, rpm = 280, d = 8 in., μ = 0.02, ΔT = 300 °F, k = 20 BTU/ft h °F, v = 1 ft/s, with the resulting adjusted heat flow path A/L = 0.4 ft to achieve 300 °F.

The value for ΔT = 300 °F was selected because oil will break down and flash with a higher temperature rise under normal operation.

Assuming no lubrication and increasing the friction coefficient to μ = 0.1 since the surface will be roughened, results in $\Delta T \approx 1,500$ °F, which is near the softening point of steel. This temperature will continue to increase as the surface degrades until it plastically deforms and stops transmitting torque.

The specific heat equation can be used to approximate how long (t_h) it would take the gear teeth to be raised by this ΔT. It's assumed only the gear teeth (W_{gear}) are heated quickly and then the heat flows into the shafts. This is a fair approximation since the shafts were brownish which would have been temperatures of less than 500 °F and visually on reference [4]. It also suggests a reason the bearings were still able to spin since less heat got to them.

The heat rate gained due to the temperature difference in a given period of time (t_h) is:

$$Q_{in} = W_{gear}{}^*C_{steel}{}^*(T_{final} - T_{initial})/t_h \text{ Btu/h}$$

$$t_h = W_{gear}{}^*C_{steel}{}^*(\Delta T)/Q_{in}$$

Using W_{gear} = 115 lb, C_{steel} = 0.1 Btu/lb °F, ΔT = 1,500 °F, Q_{in} = 71,000 BTU/h (Eq. 1)

$$t_h = 0.25 \text{ h}$$

This shows that a loss of oil can be expected to have a catastrophic effect on the gear teeth in less than an hour under full load conditions.

The cause of why the oil supply might have been lost should be investigated further. Even in an unloaded condition the gear distress might not have been as severe but major gear damage could still have been expected.

That's the benefit of these simple closed-form solutions. They aren't that accurate but do provide an easy to explain and immediate answer. Even if a more elaborate analysis was performed, many assumptions would still have to be made and the conclusions would have probably been the same. It's a very useful way to screen failures.

References

1 Akbarzadeh, S. (2010). Elastohydrodynamic analysis of spur gears using load-sharing concept: running-in and steady state, LSU Doctoral Dissertation.

2 Errichello, R. (2021). *The Lubrication of Gears – Part III*, https://www
.geartechnology.com/issues/0791x/errichello.pdf.

3 Sofronas, A. (2006). *Analytical Troubleshooting of Process Machinery and
Pressure Vessels: Including Real-World Case Studies*. Wiley.

4 Handschuh, R., F., Gargaon, L. J. (2021.). Test facility simulation results for
aerospace loss-of-lubrication of spur gears. *NASA/TM-2014-218396*.

3.24 Analyzing the Total Collapse of a Multi-Story Building

A portion of my career has involved investigating the causes of catastrophic type failures on machines, pressure vessels, and structures in industry. Causes and solutions are required quickly so the equipment can be safely restarted and production resumed.

While collecting and analyzing the damaged and mangled parts, my approach is to propose and test a theory on the most probable cause. Some form of an analysis is used based on the available data. I share this information with the investigating team and modify this as new information is obtained. This helps focus the team on all the confusing debris and input from observers and helps in understanding what might have occurred. Testing the theory in this way will either validate it or provide reasons to dismiss it and concentrate on another theory. This is similar to what is done using the scientific method [1].

This approach is different than speculating on a cause. Speculating, which has its merits, is saying what you think may have happened but not testing your idea against the data that is available. You are not trying to prove or disprove your idea.

Speculating can have a negative effect on a failure investigation, especially when it's done by a recognized expert. Key items may be overlooked. Without the true causes revealed by reviewing and testing all of the data, the correct remedial actions may not be implemented.

Consider theorizing on the collapse of the Champlain Tower South, Surfside, Florida in June, 2021 [2]. The following information was available from many internet sources, such as reference [3] about a month after the failure. Historically for a catastrophe such as this a final failure report may take 2 or 3 years to be issued.

(1) A video of the pancake-type collapse of the building was available [4].
(2) A sketch of the building and views of the sections that collapsed were available.
(3) The slab section next to the swimming pool had dropped 10 ft into the underground garage as seen on photos. The swimming pool was a rigid structure that went to the garage floor. It was not a contributor to the failure.

(4) Several persons in the building reported seeing a section of the pool patio slab fall into the underground garage several seconds before the building collapsed.
(5) There were inspection reports of cracking and heavy corrosion in the underground garage section beams and roof where the failure occurred.
(6) The initial dust cloud of the collapse appears to come from the first floor.
(7) There was a delay in the time for the three sections of the building to collapse.
(8) Deck core samples take in 2018 cause the concrete strength to be questionable.
(9) Punch-out shear of parking garage roof slab suggests the slab dropped vertically.
(10) The underground garage was also under the building. The slab that was seen sinking was not under the building.
(11) Questions remain on why the slab dropped as it did. Corrosion and cracking of the columns could have weakened the concrete to the point it just slowly failed and that was the trigger.

A preliminary theory is that when the pool deck/garage roof slab dropped due to deteriorated concrete, the weight of the slab initially pulled on building columns with tensile force (T) as shown in Figure 3.24.1. It did this through the concrete and rebar connection as shown in Figure 3.24.2. When the first row had collapsed it moved to the next column row. Progressive failures of the columns occurred because of the weakened structure falling into the underground garage. This continued to progress to the point that pancaking of the building into the underground garage occurred, meaning one floor drops on another, started from the bottom floor. The building came down straight because of the support of the shear wall on the building which remained standing. The cross-hatched section in back of the shaded section was the second and third to collapse several seconds later. Some say the first and second came down together as it is difficult to tell from the video [4]. Only the unshaded portion of the building away from the collapsed slab section remained standing.

It is first very important to understand how corrosion and concrete degradation on the supports under the failed section could have caused the garage roof to weaken. First, the original design loads will be approximated and compared with what the concrete strength would have had to deteriorate to so it would fail. It is assumed the concrete will be compression tested as part of the investigation.

The weight (W) of the failed slab is as follow and the slab and columns are assumed to be 1 ft thick and 1 ft^2, respectively. These dimensions will have to be verified.

W = fallen slab area*slab thickness*concrete-rebar density

$= 5,800 \text{ ft}^{2*}1 \text{ ft}^*150 \text{ lb}_f/\text{ft}^3 = 870,000 \text{ lb}_f$

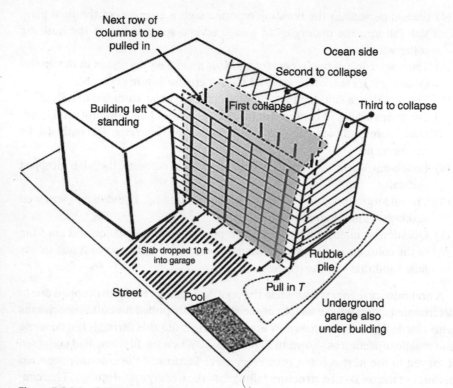

Figure 3.24.1 Building collapse scenario.

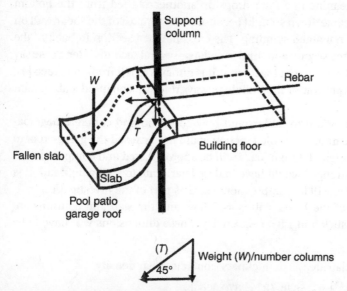

$(T) = (\text{Weight } (W)/\text{number columns})/\sin(45°)$

Figure 3.24.2 Development of lateral force (T).

Consider the compressive stress (σ_c) on a column that collapsed and the punch-out shear (σ_{po}) of the slab that was shown in photos. Shear strength will be approximated as 10% of compressive strength for concrete.

σ_c = W/number garage support columns/

$\qquad$ × (144*compression area column ft^2)

$\qquad$ = $W/8/(144*1)$ = 800 lb/in.^{2}

$\qquad$ × new failure strength compression 4,000 lb/in.^{2}

σ_{po} = Weight slab/number columns/(144*perimeter area column ft^2)

$\qquad$ = $W/8/(144*1*1*4)$ = 190 lb/in.^{2}

$\qquad$ × new failure strength shear 400 lb/in.^{2}

So from these crude calculations, it appears that if the compressive strength is measured as half of what it originally was, meaning 2,000 lb/in.^{2} a shear type failure, meaning punch-out failure, would be possible. This would suggest water soaking, cracking and rebar corrosion of the concrete under the deck was a real possibility of weakening the structure to a collapse condition.

The following describes how this dropped slab could have caused the building to collapse from the lateral force (T).

Consider that the dropped slab section fell straight down and it was tied into the building's columns with rebar. Figure 3.24.2 shows that as it fell the weight of a portion of the slab section pulled on the rebar and column with the assumed 45° angle. There would be no load (i.e. $T = 0$) on the rebar if all supports were present under the slab, as originally designed.

The amount of the dropped slab weight (W) which cause (T) is unknown as it depends on the edge support. Assume for now that all the supports under the slab failed and only the one side edge attached to the building was still connected.

$\qquad T = [W/\text{number of columns}]/\sin(\theta°)$

$\qquad T = 1.414*W/5 = 246,000 \text{ lb}_f$

As a point of reference on the order of magnitude, a 55 ton tracked bulldozer, like a Cat D9, could push on each column with a force of about 160,000 lb$_f$.

The pancaking collapse of the building is clearly shown on the video. Looking at Figure 3.24.3 assume (s_L) is suddenly removed when all of the column supports collapse.

The weight above is $\sum W_{2-10}$.

The structure will drop a distance (s) with an acceleration of gravity (g) in (t) seconds:

$\qquad s = 1/2*g*t^2$

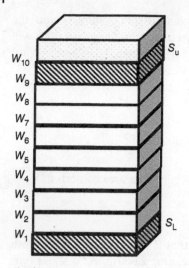

Figure 3.24.3 Multi-story building with floor removed suddenly.

$t = [2*s/(32.2)]^{1/2} = 0.7$ seconds per each crush depth 8 ft floor.

This is much faster than what has probably occurred as shear drag would have slowed things down.

The impact force will be $F = m*a = \sum W_{2-10}*a/g = \sum[W_{2-10}/32.2]*[2*s/t^2)]$

An impact force $\approx 2*\sum W_{2-10}$ lb$_f$ is usually used on structures. This would be a time of $t = 0.5$ seconds per floor collapse.

For this analysis, it doesn't really matter which floor collapsed first as the weight above would probably collapse the next even if it was the second from the top (s_u). Most buildings are designed with a load factor of 2–3. Presently there is not much convincing evidence that the failure originated at the top.

Since the cloud of smoke from the debris appears to start from the bottom (s_L) not the top (s_u) something happened to cause loss of support at (s_L). It is assumed to be the slab, meaning the top of garage, collapse as described.

It has been my experience that several events usually occur to result in a failure. Lack of or delayed maintenance, overloads due to exceeding operating or design specifications, unauthorized repairs, or modifications are typically the causes. Major design deficiencies from the manufacturer are quite rare if built to their specifications and if these aren't altered to save time or reduce costs. A long history of successful operation before a failure certainly would direct attention to these possibilities.

That may also be the case in this building collapse as it's unlikely only one or two defective columns will bring a multi-story building down. Degradation of the supporting structure by corrosion and cracking, non-code design modifications or poor design or construction techniques in the original structure or a combination of these may have occurred. The exact triggering event is unknown.

Continued review of the data such as concrete testing of the slab and supporting columns will help determine if this theory still has any merit.

This failure was very difficult to research as every picture reminded me of the 95 deaths and 14 missing as of 14 July 2021. Many families and individuals lost their lives in the collapse. May God continue to look over them and their grieving families and friends.

References

1 Wikipedia. (2021). Scientific Method.
2 Wikipedia. (2021). Surfside Condominium Building Collapse.
3 Ostroff, J. Here's Cause of Miami Condo Collapse Champlain Condo Towers, Surfside, First of series, July, 2, 2021. YouTube.
4 Lee, W. (2021). Surveillance video shows the moment when Surfside condo collapses. Associated Press (31 June 2021).

4

Fluid Flow and Heat Transfer Examples

Saying You Always Find The True Cause Means You Haven't Analyzed Many Failures

4.1 Addressing Heat Exchanger Tube Leaks

Heat exchangers have multiple tubes with a fluid through and over them which either heats or cools another fluid. Tube leaks may not be cost-effective to retube and common practice is to plug the leaking tubes stopping the flow through them. This will also change the process temperature. What can an engineer contribute when asked if the 20 tubes plugged in a single-pass 325 tube exchanger are too many?

Design programs aren't always available to analyze the situation when an immediate answer is required. Removing and shopping the exchanger would cause a long delay and the unit was due for a complete retubing in two years. Some were saying that plugging 10% of the tubes was acceptable.

There are important factors to consider. One is that the velocity in the remaining tubes shouldn't increase so much that erosion becomes a concern. Another is that the heat transfer process shouldn't be unfavorably diminished. Also after observing the plugged tubes, the question asked was if the tube and tube sheet stress would increase excessively.

Three equations will be used to help make a quantitative educated decision, which are the continuity equation, the convection heat equation, and a thermal stress equation.

In this exchanger, each tube is designed for a liquid flow velocity (V_{tube}) of 7 ft/s. Reducing flow below 3 ft/s could cause fouling and too much higher in excess of 15 ft/s and erosion might occur. It depends on the tube material and fluid cleanliness. Tube vibration is not considered here.

Unique Methods for Analyzing Failures and Catastrophic Events: A Practical Guide for Engineers, First Edition. Anthony Sofronas.
© 2022 John Wiley & Sons, Inc. Published 2022 by John Wiley & Sons, Inc.

The weight flows through each tube where $Q = V_{tube}*A_{tube}$ ft^3/s:

$$W_{tube} = \rho*A_{tube}*V_{tube} \text{ lb/s}$$

In the equation, ρ is fluid density, A_{tube} is tube flow area, and V_{tube} is the velocity through the tubes, all considered constant for this evaluation.

The convection heat equation represents heat transferred (q) from a heat exchanger. A convenient form for flows of liquids is:

$$q = U*A*F*T_m \text{ Btu/h}$$

where q is the heat dissipated by the heat exchanger. The heat transfer constants (U), correction factor (F), and the logarithmic mean temperature difference (T_m) will be considered constant. The surface area of a tube is A (ft^2).

Keeping the weight flow the same but reducing the number of tubes increases the velocity in the unplugged tubes and also reduces the heat flow since the tube surface area (A) is reduced.

This means that for an estimate this is proportional to the number of tubes plugged.

The velocity in a tube is:

$$V_{plugged} = V_{unplugged}*N_{tubes}/[N_{tubes} - N_{tubesplugged}]$$

Likewise for the effect on the heat transferred:

$$q_{plugged} = q_{unplugged}*[N_{tubes} - N_{tubesplugged}]/N_{tubes}$$

With a 325-tube exchanger and 20 tubes plugged (6%),

$$V_{plugged} = 7*1.06 \approx 8 \text{ ft/s or } 6\% \text{ increase in velocity}$$

$$q_{plugged} = q_{unplugged} \text{ or } 6\% \text{ less heat transferred}$$

One other concern on the plugging of tubes is the effect the temperature difference has on this group of tubes and the force on the tube sheet. Since the shell and the tubes grow at about the same amount, the cooler plugged tubes will have a higher stress since they don't grow as much. They are stretched and produce a new load on the tube sheet and tubes.

A simple equation will be useful here.

The force is on one tube (F_{tube}) of restrained parallel steel alloy tubes of equal length but different temperatures:

$F_{tube} = 200*A_{section}*[T_{bulk} - T_{tube}]$ lb$_f$ where ($A_{section}$ in.^{2}) is the cross-sectional area of the tube.

When one isolated plugged tube is at 400 °F temperature and the rest of the tubes are at the design temperature of 500 °F and $A_{section} = 0.33$ in.^{2},

$$F_{tube} \approx 6,600 \text{ lb}_f$$

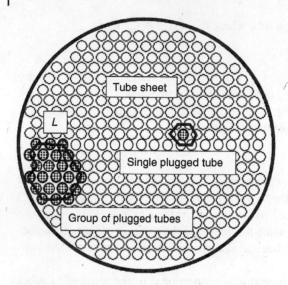

Figure 4.1.1 Grouped and isolated plugged tubes.

Load on the tube sheet from a group of tubes,

$$F_{\text{tubesheet}} = F_{\text{tube}} * N_{\text{plugged}}$$

The shear stress (S_{ss}) on the tube sheet is approximately:

$$S_{\text{ss}} = F_{\text{tubesheet}} / [L * t_{\text{ts}}]$$

where t_{ts} (in.) is the tube sheet thickness and L (in.) is the measured metal ligament length between tubes (Figure 4.1.1).

This 325-tube exchanger had 20 closely grouped tubes plugged as shown in Figure 4.1.1. This can be compared with an isolated tube.

The shear stress in the tube sheet on the perimeter (L) for the grouped case was about $S_{\text{ss}} = 9,000$ lb/in.^{2} and for the isolated tube $S_{\text{ss}} = 1,000$ lb/in.^{2}. This stress level is much higher than an isolated plugged tube but probably wouldn't cause a catastrophic type failure. This is because the small displacements and stresses would be limited by yielding or tube flexing.

The maintenance team reviewed the data and decided to retube this exchanger early because of the heavy pitting.

The rule of thumb that says up to 10% plugged tubes is acceptable is shown to have merit. It is important that a detailed analysis on the specific case should be considered in the decision-making process when safety is a concern.

4.1.1 Summary

Plugging 10% of leaking tubes in heat exchangers seems to be acceptable and not to cause heat transfer issues.

4.2 Explaining Flow Through Piping Using the Poiseuille Equation

This experimental equation gives the pressure drop in a liquid fluid in laminar flow, flowing through a long cylindrical pipe of constant cross section. The pipe length (L) is considered much larger than the radius (r). The equation is called Poiseuille's law after the French scientist J. L. Poiseuille (1799–1869), who derived it in an attempt to understand the flow of blood.

$$\Delta P = (8*\eta*L*Q)/(\pi*r^4)$$

ΔP is the pressure difference between the two ends, L is the length of pipe, η is the viscosity, Q is the flow rate, and r is the pipe radius.

The resistance to laminar flow of an incompressible fluid having viscosity through a horizontal tube of uniform radius and length is given by:

$$R = 8*\eta*L/(\pi*r^4)$$

The flow in the pipe is given by:

$$Q = \Delta p/R$$

The use of this equation is used in describing the flow of blood through the body in Section 8.10.

An actual example of its use in industry was to explain what had occurred during a bearing failure in a large diesel engine. The question was if any damage may have been done to the other bearings.

Consider Figure 4.2.1 where a main bearing has been damaged increasing the clearance in the bearing thus allowing more flow (q_2).

All that needs to be done is to look at the Poiseuille equation as no numerical answer is required, only an explanation.

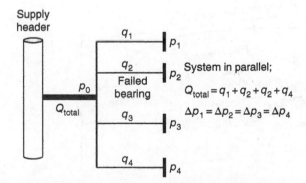

Figure 4.2.1   Bearing failure in diesel main bearing.

With a decrease in R, where $R = 8*\eta*L/(\pi*r^4)$, due to the clearance increase that acts as a large orifice or radius, there will be an increase in flow in q_2 since $q_2 = \Delta p_2/R$.

With an increase in flow in the failed bearing and Q_{total} remaining the same, the other bearings will see less flow since more flow is seen at the failed bearing. Thus, it would be a good idea to flush the system, change the damaged bearing, and check the other bearings either by disassembling them and looking at them or by jacking them.

4.2.1 Summary

When a simple explanation of the flow through a pipe with a small diameter in relation to its length is required, the Poiseuille equation can help. It can show the effect of pressure drop and viscosity of the liquid and how it relates to flow.

4.3 A Local Flooding Event at a Plant Site

During a heavy rainstorm flooding occurred in a plant at a location that hadn't flooded before, concern was that the underground storm sewer lines had become plugged with debris. There are several miles of these underground lines. No one wanted to have to flush out all of these lines, so the question was if it might have been due to the 6 in. of rain that fell in an hour (iph). It may have been too much for the storm sewer drains to handle. Such rains are rare and the amount of flooding that occurred hadn't happened in the past. Typically, a heavy rain is 2 iph for the same time period. The question asked was if the drains were plugged or was it just due to the sudden high rainfall rate?

An analysis can help answer this question. Figure 4.3.1 shows a depressed basin-type area (A_{basin}) with the storm sewer drains that had flooded.

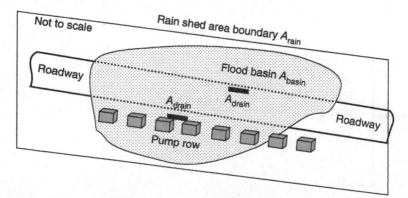

Figure 4.3.1 Flooded basin area.

The surface area (A_{rain} ft^2) is the area that captures the rainfall and directs it to the drains and is called the rain shed area. When the rainwater height (h ft) in the basin (A_{basin}), which is a low point, reaches the height of the curb drain, the water starts to build up. The drains are gravity-type drains with a flow area of (A_{drain} ft^2) and there are (N) numbers of them to handle the basin area. The rainfall rate is r ft/s over a period of t_s. With this information, an analytical model can be built. The theory is much like filling up a bathtub. When the water going in, meaning the faucet or rainfall from the rain shed, is greater than the water going out, meaning the drain or storm sewers, the level (h ft) will build up in the basin and eventually fill to height (h ft) in (t_s) seconds.

The water going into the basin in cubic ft per second (cfs) is:

$$Q_{in} = A_{rain}*r$$

The water going out of the drains due to gravity alone with no mechanical pumps is [1].

$$Q_{out} = 0.8*N*A_{drain}*(2*g*h)^{0.5}$$

The difference is the amount stored that builds up with time (t_s).

$$(Q_{in} - Q_{out}) = Q_{stored} = A_{basin}*h/t_s$$

$$A_{rain}*r - 0.8*N*A_{drain}*(2*g*h)^{0.5} = A_{basin}*h/t_s$$

This equation can be solved for h by iterative or closed-form methods.

During the storm $N = 2$, $A_{drain} = 2.0$ ft^2, $A_{rain} = 320,000$ ft^2, $A_{basin} = 25,000$ ft^2, $r = 1.4*10^{-4}$ ft/s (6 in./h), $g = 32.2$ ft/s^2, $t_s = 3,600$ seconds.

In Table 4.3.1, the results show that the drains just couldn't handle 6 iph of rain and the level built up during the rainfall. A normal heavy rainfall of 2 iph doesn't appear to build up above the drain height.

Some idea if the drains were plugged can be determined by noticing how long it takes for the basin to empty after the rain had stopped.

$$Q_{drain} = 0.8*N*A_{drain}*(2*g*(h/2))^{0.5} = 23.7 \text{ cfs}$$

(t) time to empty = $A_{basin}*h/Q_{drain} = 1,795$ seconds or 0.5 hours

The site was dry again in less than an hour meaning it had drained at the capacity it should be due to gravity alone. This suggests that the drains were acting as

Table 4.3.1 Flow model results.

Condition	Flow entering Q_{enter} (cfs)	Flow leaving Q_{leave} (cfs)	Result (h ft)
Heavy rain 6 iph	44.4	32.5	1.6 ft build up
Normal rainfall 2 iph	14.8	12.8	0.3 ft build up

designed and probably weren't plugged. It appears that the heavy hourly rainfall was what had caused the flooding.

This information was transferred to the plant facilities manager so that a plan could be developed for any future sudden heavy rainfalls of this magnitude.

4.3.1 Summary

This type of analysis can be used to analyze drainage and flooding problems.

Reference

1 Sofronas, A. (2006). Analytical troubleshooting of process machinery and pressure vessels. In: *Including Real-World Case Studies*, 142. Wiley.

4.4 Examining Fan System Pulsations

This case came up several years ago when working as a consultant on an environmental control system. This system took a contaminated gas and ran it through a large scrubber. While the problem was with the scrubbing unit, the project supervisor felt the problem was with the large fan supplying the air to the unit. He said there were pressure variations to the vessel it was supplying the air too and also a low-frequency vibration in the ductwork. He said it seemed like a couple of cycles per second and asked if I had any ideas on what the problem might be. He supplied some operating pressure data and a sketch of the system ductwork and a fan curve. A visit to the unit showed that there was the low vibration mentioned. There was a lot of other vibration also present such as the sloshing forces of the mixers in the large 120-ft diameter stirred tank.

This is typical for consultants or most problem solvers. When gathering data, there will be many who feel they have the answer. Their inputs need to be examined and explained to those involved, as is done here.

Figures 4.4.1 and 4.4.2 show typical data provided.

The problem at first sounds like it might be an instability of some kind. A fan blowing air through ductwork of a small volume into a large volume vessel can act as a resonating chamber. In technical terms, this system is called a Helmholtz resonator and is usually used as a tuning tool to reduce pulsations.

A crude calculation of the vibrating air-mass natural frequency (f_n) in the vessel for such a resonator is as follows when the following is true [1] (Table 4.4.1):

$$\text{Is } L > (1/2)*(\pi*A)^{1/2} = 22 \text{ ft OK}$$

$$f_n = (C/2\pi)*[A/(L*V)]^{1/2} \text{ cycles/s (cps)}$$

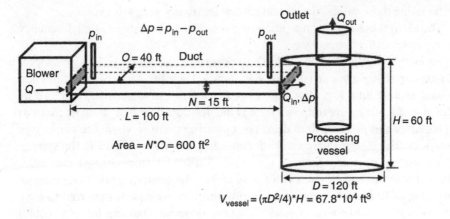

Figure 4.4.1 Blower vessel system.

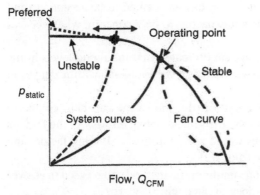

Figure 4.4.2 Fan curve.

Table 4.4.1 Terms in frequency equation.

Term	Value	Significance
C	1,374 ft/s	Speed of sound in air
V	$67.8*10^4$ ft^3	Resonator volume
L	100 ft	Length of duct
A	600 ft^2	Cross section of duct

Inserting these values, the system natural frequency is $f_n = 0.7$ cps.

The air in the ductwork and processing vessel had a pressure cycling frequency of Δp around this frequency.

To further verify instability wasn't a concern, a system curve was developed. To explain this, only a generic curve is shown. Here are some of the things you would explore on a Fan Map. Figure 4.4.2 is a Fan Curve normally supplied by the manufacturer. Superimposed on it is the developed system curve. It contains geometries and pressure drop data. The operating point is where the two curves intersect. The system curve can shift depending on the resistance in the system. For example, pinching back on a valve or fouling of an exchanger can move the point to the left. When a Fan Curve is flat, the operating point can become unstable as it doesn't know where to run and can cycle. It is preferred to have an upslope as shown so no unstable conditions occur. There's a lot more useful information available on these types of curves and they are a good analysis tool to use.

After reviewing this data, the fan system design seemed to be acceptable and the low-frequency disturbance was not expected to be a concern, whatever was the true cause.

Further review of the processing system indicated that the liquid levels in the processing vessel and its dynamics were causing the problems within the vessel and not the fan system.

This shows that a few equations and a little knowledge can help engineers reduce the number of unknowns in their troubleshooting efforts. As the famous fictitious detective Sherlock Holmes would say, "Eliminate all other factors and the one which remains must be the truth" [2].

As is usually the case, the use of one mathematic method can be used to answer other questions. For example, how does a police whistle work?

You know the whistle type. It has a pea or something in it and when you blow in it, it produces a high-pitch trilling tone. Referees in sports use them and we all had one to annoy people with as kids. Figure 4.4.3 shows a typical one.

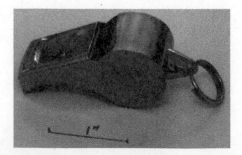

Figure 4.4.3 Police-type whistle.

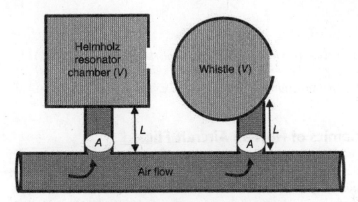

Figure 4.4.4   Helmholtz and whistle.

I would have thought the mechanics very complicated, but since having worked with vibration of systems, the following analysis was intriguing. When air is blown across the top of a soda bottle, it whistles and the air inside vibrates and causes a tone.

This action is like a Helmholtz Resonator. Basically what happens is the air in the bottle is springy. The volume of air acts as a mass and a spring and vibrates at its natural frequency or resonates. The similarity of both is shown in Figure 4.4.4.

This natural frequency is easily calculated using the formula:

$$f_n = (C/2\pi)*(A/(L*V)^{1/2} \text{ cycles per second (cps)}$$

where $C = 1{,}374$ ft/s or the speed of sound in air, A is the area shown, V is the volume, and L is the neck length.

For the whistle, the round part diameter is 0.9 in., $L = 0.75$ in., its width is 0.7 in. and the area of the rectangular slot to blow into 0.2 in. by 0.7 in.

Since all values are in the ft lb s system,

$$V = 0.000258 \text{ ft}^3, A = 0.00097 \text{ ft}^2, L = 0.063 \text{ ft}$$

$$f_n = 1{,}690 \text{ cps}$$

The sound of the whistle has a frequency of about 2,200 cps, when checked against a tone generator. So the analysis is off some but close enough to help understand the physics of blowing that whistle. Holding the pea inside and a flat pure tone is produced, so the pea is probably what produces the trill sound.

4.4.1 Summary

Vibrations can be developed by the flow of air into a chamber. The air in the chamber vibrates at its natural frequency and develops an excitation force based on the air-mass dynamics.

References

1 Sofronas, A. (2012). *Case Histories in Vibration Analysis And Metal Fatigue*, 134. Wiley.
2 Doyle Conan, A., *The Adventures of Sherlock Holmes, The Sign of Four*, 1890.

4.5 The Dynamics of How an Aircraft Flies

There are several theories on this, all of which are true to a degree and the theory can get quite complex. Bernoulli's principle considers the lift due to lower pressure with faster flow over the top of the wing causing the lift. All I know is that if I put my hand out the window of a moving car, my hand tries to fly as shown in Figure 4.5.1. The same thing happens running with a kite. This is described as the Newton Equation Solution and both methods are complementary to each other.

As the hand is put into the air stream, it moves back and up and the car speed provides the energy. No hand angle means parallel to the road and it doesn't lift. Let's see how this can be modeled.

Consider a flat airfoil deflecting air out of its path moving to the left (Figure 4.5.2).

Its velocity is V_o and the volume of air it moves is Q_o. When it is pushed by the airfoil at an angle $\theta°$, the air is deflected downward as shown. Due to continuity, the velocity entering and exiting are about the same.

$$Q_o = w_{wing}*L*\sin \theta°*V_o = A_{total}*\sin \theta°*V_o \text{ ft}^3/\text{s}$$

Multiplying by density:

$$F = (\rho/g)*Q_o*V_o \text{ lb}_f$$

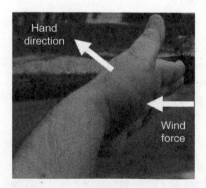

Figure 4.5.1 Flying hand out car window.

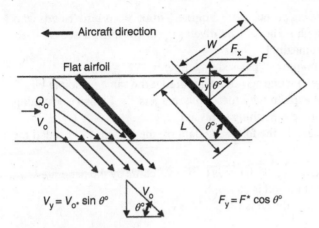

Figure 4.5.2 Airfoil lift.

Geometry for the lift component only (F_y) will be analyzed since the other component (F_x) is drag. The power plant must produce enough thrust to overcome both.

$$F_y = F^* \cos \theta° = (\rho/g)^* Q_0^* \cos \theta°^* V_0$$

Making the substitution:

$$F_y = (\rho/g)^* A_{total}^* (\sin \theta°)^* (\cos \theta°)^* V_0^2$$

Notice the area (A_{total}) is used in place of the wing area $(w_{wing}^* L)$. This is because the inclined underbelly also acts as a lifting body.

My experience is that my 2,300 lb Cessna 172 flies off the runway at an angle of attack (θ) of 18°, which includes the built-in wing twist angle with an airspeed of 70 mi/h (103 ft/s).

Consider that all underbelly surfaces, including the wing act as a lifting body with a total area of $(A_{total}) = 240\,\text{ft}^2$, $\rho = 0.078\,\text{lb/ft}^3$, $g = 32.2\,\text{ft/s}^2$.

$$F_y = 1{,}800\,\text{lb}$$

This is 78% of the lift force required to lift the weight of the aircraft. This is close enough to satisfy my curiosity.

Thus the wing, tail, and underbody of the aircraft are pushed up due to plowing the air down and producing a higher pressure under the wing than on top of the wing, forcing it up. There are other factors that produce the rest of the lift necessary to overcome the 2,300-lb weight.

Since similar aircraft with more horsepower and strength can fly upside down with the correct angle of attack, this helps explain why this is so. As they say, "With enough power you can fly a brick."

With aircraft you have a big propeller and engine. Think of our bird friends who are able to take off, land, glide, hover, and dive at up to 200 mi/h based on their body power alone. Truly amazing!

The same analysis technique can be used on a Boeing 777.

I flew in a Boeing 777 jet a while ago and wondered if the same principle I used on my Cessna 172 could explain why these massive jets fly. I wasn't designing anything but wanted to prove something to myself.

After a short Internet search, the following data on the Boeing 777 needed for the analysis was located:

- Aircraft maximum weigh 660,000 lb needs to be overcome by the generated lift
- Aircraft rotation speed at takeoff is 200 mph
- Angle with built-in twist at lift-off is 18°
- Wing area 4,605 ft^2
- Tail area 1,400 ft^2
- Body area 4,000 ft^2

This very simple analysis calculates about 93% or (613,000/660,000)*100 of the lift needed to overcome the gross weight of this beautiful behemoth of an aircraft during a maximum weight takeoff.

Even if the calculations aren't accurate, I do feel better looking out the window and knowing why the aircraft will eventually rise after takeoff and that's comforting.

Another interesting flight question is, "How Can Bumblebees Fly?"

Bumblebees are interesting insects and look too bulky to be able to fly. Proving that they can would explain how other creatures fly. They might look different but the basic principles probably still hold true.

The method used on wing lift will utilize the equations developed for aircraft. They will not be repeated here since they should already be on a spreadsheet if the reader has interest in the analysis part.

Consider the bee in Figure 4.5.3. Just like a propeller it has wings that beat up and down at some speed V. When they beat down, they displace air causing lift. The pressure under is higher than the pressure on top. When the wing goes back up, it also rotates to reduce the pressure on top, as slow-motion videos clearly show. If this didn't happen, the force up would cancel the force down and it would just bounce around.

The little bee can do a lot of other clever things with its wings to make them go forward, up and down and turn. They can go backward too as well as hover and glide. Let's just consider getting the bee airborne.

First some bumblebee data for the spread sheet:

- Length (L) = 1/2 in. (0.04 ft)
- Width two wings ($2*w$) = 2 in. (0.17 ft)
- Beats per second = 230

Figure 4.5.3 Bumblebee flight.

- Down stroke = 1 in. (0.08 ft)
- Weight 300 mg (0.0007 lb)
- Wing angle assume optimum at 40°
- Area wing = 0.007 ft²

Getting beats per second into velocity:
Time for a beat = 1/250 = 0.004 s
Velocity = distance(stroke)/time = 0.08/0.004 = 20 ft/s
Putting this information into a spread sheet:
Lift force F_y = 0.003 lb and since the weight of bees is only 0.0007 lb, it should fly with no problems even carrying a sizable pollen load on its legs.

Hummingbirds are larger and their wings beat slower at 80 beats per second and have a larger wingspan. The principle of flight is similar to that of the bumblebee but the bumblebee is heavier relative to its wing size so its wings have to beat faster at 230 beats per second.

4.5.1 Summary

The same lift equations for small aircraft work for large aircraft. Pressure under the wing and body due to the forward motion, which is higher than the pressure on top, pushes the aircraft up. This type of flow is also useful when explaining the flow in mixers, pumps, and compressors. Bumblebees can fly because of their unique wings that beat extremely fast and can pivot in a unique way for maneuvering and drag resistance.

4.6 How Much Wind Does It Take to Blow Over a Motor Coach?

I was on vacation with my wife and there was a strong crosswind. My pickup was affected by the gusts. The weather report said the gusts were 50 mph. I noticed

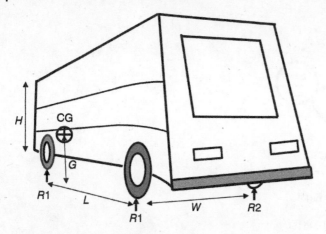

Figure 4.6.1 Motorcoach dimensions for FBD.

that motorcoaches were having a difficult time keeping straight down the road and were swerving. I wondered how much of a gust of wind was required to roll one over. A force balance could help answer this question. Consider the sketch of a motorcoach in Figure 4.6.1 with the dimensions and center of gravity (CG) shown.

For this case, let's assume the wind is acting on the area $A = L*H$ and it all can be considered to act at the center of gravity (CG) at height (G). The reactions on the ground are R1 for the front and back tires to the road on the wind side and R2 for the front and back tires for the opposite side. The R2 side is the side the coach would roll over on.

Now the free-body diagram (FBD) is shown in Figure 4.6.2. The force of a gust of wind on an area is also shown.

$$F_{wind} = 0.005*A_{sqft}*mph^2$$

Summing the moments about R2 where W is the coach width:

$$F_{wind}*G - R1*W = 0$$

$$R1 = F_{wind}*G/W = (0.005*(H*L)*mph^2)*G/W$$

The reaction on the wind side two tires trying to hold the coach down without any wind is R1 = (weight of coach)/2 and when this reaction ($F_{reaction}$) becomes zero, it will be assumed the coach will blow over.

$$F_{reaction} = (weight of coach)/2 - (0.005*(H*L)*mph^2)*G/W$$

Consider the weight of the coach = 30,000 lb so R1 = 15,000 lb, H = 13 ft, L = 43 ft, W = 8 ft, and G = 5.0 ft.

For a wind gust of 90 mph, $F_{reaction} \approx 1,500$ lb that is probably enough for the R1 wheels to lift off the ground and the coach to roll over.

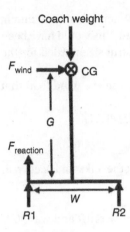

Figure 4.6.2 Motorcoach force FBD.

4.6.1 Summary

It would take pretty strong gusts of wind, around 90 mph, to blow over a motor-coach. Loss of control and swerving of the coach at lower wind speeds would probably be the deciding cause of the coach rolling over. This is a wind force of about 23,000 lb distributed over the side of the coach.

4.7 How Much Wind Force to Buckle an Aircraft Hanger Door?

The door of an aircraft hangar similar to Figure 4.7.1 had bucked in after a heavy wind. Observing the failure, it was the angle iron struts braced against a beam used to hold the door that had buckled. The folks repairing the door said they

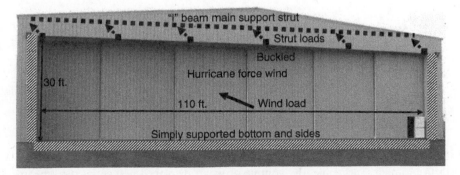

Figure 4.7.1 Aircraft hanger door subjected to hurricane-force winds.

were going to increase the size of the angle iron struts. When asked how much, they said about double the size. Since a failure had occurred, this would have been the place for a few simple calculations to determine the strut size needed for the worst-case scenario.

Figure 4.7.1 also shows how the doors were supported and the wind load that was on the outside surface.

The pressure on an area for a wind of velocity in mi/h (MPH) is:

$$p \, \text{lb/ft}^2 = 0.005 * \text{mph}^2$$

For a hurricane-force wind gust of 75 mph which is what the folks said occurred:

$$p = 28 \, \text{lb/ft}^2$$

The door was supported top and bottom and had a span of 110 ft and was 30-ft high as shown in Figure 4.7.1.

The wind load was therefore:

$F_{door} = 28*3,300 = 92,400 \, \text{lb}$ and the top struts which are attached to the door and the main "I" beam experience half of this and there were 6 struts.

$F_{strut} = 92,400/(2*6) = 7,700 \, \text{lb}$ that each strut is required to support without buckling. This is also true for other support points for this load.

The critical bucking load is:

$P_{critical} = \pi^2 * E * I/(4*L^2) = 5,200 \, \text{lb}$ and this explains the buckling since the wind load on a strut is greater than this.

For the above case, $I = 1.2 \, \text{in.}^4$ for $3 \times 3 \times 1/4$ in. angle, $L = 120$ in. length and $E = 30*10^6 \, \text{lb/in.}^2$

Think of what this would be with a 120 mi/h hurricane. The force on each strut would increase by a factor of 2.6.

It would be wise not to guess and design for about a 16,000-lb force per strut and use tubular struts that are resistant to buckling. With 120 mph wind, the roof will probably be gone and the door will fail from some other cause, meaning no roof for support. For lower wind speeds, you would have solved the problem.

4.7.1 Summary

For a big door not well supported, the wind force of 75 mph is enough to cause permanent damage.

4.8 How Much Water on a Road to Float a Car?

During recent floods, I wondered how much water across a road would float a car. By Archimedes' principle, the volume of water displaced will equal the upward thrust on the body. That's why steel hull ships float and also concrete canoes.

Consider a car of volume $V = A*d$, where d is the depth of the car in water, and A is the car width times length considered as 5 ft*12 ft.

With water weighing $\rho = 62$ lb/ft^3 and a car weighing 4,000 lb, it will float away when:

$A*d*\rho = 4,000$ lb

$d = 4,000/(60*62) = 1.1$ ft

Considering a car bottom at 1 ft, the depth to float would be about 2 ft of water.

This is a good reason not to drive across flowing water when you don't know the depth.

4.8.1 Summary

A couple of feet of water is enough to float a car. With flowing water, it could float the car away.

4.9 How Fast Does an Object Hit the Ground?

An article in a newspaper stated that a piece of an aircraft had broken loose from about 20,000 ft. I've always wondered what speed objects achieve before they hit the ground.

The maximum speed an object achieves when it free falls from a given height is called its terminal velocity ($V_{terminal}$). This is the velocity it obtains when the force of gravity pulling the object toward Earth is just balanced by the air resistance which is causing it to slow down. The math to show this is simply performing this balancing act and solving for the velocity:

Force balance and for equilibrium, $F =$ weight–drag:

$$F = m*g - 1/2 \, \rho*V^2*A*C_d = 0$$

Solving for (V) meaning $V_{terminal}$

$$V_{terminal} = [2*W*g/(C_d*\rho*A)]^{1/2} \text{ ft/s}$$

Air density (ρ) = 0.078 lb/ft^3; (g) = 32.2 lb ft/s^2; and drag coefficient (C_d) = 0.8 flat surface.

Table 4.9.1 shows the terminal velocity for several objects.

How hard would a baseball size piece of hail of 3-in. diameter hit the ground? Some data;

- $d = 3$ in. (0.25 ft)
- $W = 0.44$ lb
- $C_d = 0.5$

Table 4.9.1 Terminal velocity for objects.

Condition	Weight (W lb)	Area (A ft²)	$V_{terminal}$
Person free fall from aircraft laying flat	150	5	176 ft/s (120 mph)
Person free fall from aircraft head down	150	2	278 ft/s (190 mph)
Chunk of blue ice falling from aircraft	1	1	32 ft/s (22 mi/h)
Bullet shot into air	0.02	0.0007	172 ft/s (117 mph)

- $\rho_{air} = 0.078 \text{ lb/ft}^3$
- $A = 0.049 \text{ ft}^2$

$$V_{terminal} = [2*W/(C_d*(\rho_{air}/g)*A)]^{1/2}$$
$$= 122 \text{ft/s (83 mph)}$$

Assume the ground is a spring, like stepping on grass and has a spring rate of (k = 200 lb/in.) or (2,400 lb/ft). This means that if you place a 200-lb weight on the grass and the weight sinks 1 in. under the load, the spring constant is k = 200 lb/in.

Equating KE of the hail to PE of a spring meaning the lawn:

$1/2(W/g)*V^2_{terminal} = 1/2 (k*\delta^2)$ and solving for the dent depth in the lawn:

$$\delta = V_{terminal}*[(W/g)/k]^{1/2} = 0.29 \text{ ft}$$

Solving for F of impact is:

$$F = k*\delta = 697 \text{ lb}$$

This would produce quite a blow on the head and would crack a windshield. It might fracture the skull because it takes about 1,000 lb to do that.

A marble size piece of hail hitting your arm would probably leave a nice black and blue mark.

With some creative analysis, the velocity in free fall at any time (t) with drag can be determined using a hyperbolic function (i.e. tan h).

$$V(t) = V_{terminal}* \tan h(g*t/V_{terminal})$$

The graph of this is shown in Figure 4.9.1 along with the equation to be developed with a little integration and manipulation into $V(t)$.

Notice that without drag, there isn't a terminal speed and the velocity keeps increasing.

An example would be if the piece of hail with a terminal velocity of 122 ft/s hit the ground in 5 seconds:

$$V(5) = 122* \tan h(32.2*5/122) = 105.5 \text{ ft/s}$$

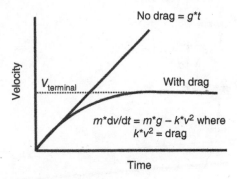

Figure 4.9.1 Free fall with drag and no drag.

4.9.1 Summary

Things dropped from high enough will have a final speed called their terminal velocity. The force it develops on a flat surface depends primarily on the weight (W) of the object, and its projected area (A). A person free falling out of an aircraft can develop a final speed of 120–190 mph and a bullet shot into the air can have a final speed of about 117 mph when it comes back down to Earth. A piece of hail the size of a marble can leave a black and blue mark.

4.10 Collapse of a Bubble and the Excitation Force on a Structure

This section develops a method that helped answer some questions. The method may not be quantitatively correct but it does provide data when none was available and allowed informed decisions to be made. This sometimes occurs when solving problems in industry.

A very large newly designed tank was used for removing contaminants from a combustion stream. The contaminated gas was bubbled through a large number of pipes into a liquid slurry to clean the gas. This is normally a processing problem and the method for doing this is confidential information. For this reason, few specifics are provided.

The reason for the interest was that these pipes were vibrating and causing excessive wear. There were several reasons for the pipes vibrating, some being due to mixers in the vessel and the others due to the gas flow through the liquid. Some on the investigating team speculated that the bubbling out of the pipes might be the most important factor forcing the pipes to vibrate. They wanted to know if reducing the flow, which would have been costly and detrimental to the process, might eliminate the vibration.

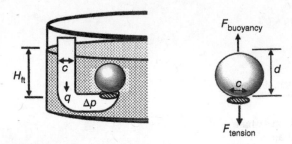

Figure 4.10.1 Making a bubble.

This simple analysis was done to help answer this question.

Figure 4.10.1 represents a section of the large-diameter tank with just one of the pipes mentioned shown.

Air is being blown into the liquid at (q) in.^{3}/s from a blower and forms a spherical bubble of (d) diameter as it exits the pipe. The pressure in the bubble is the supply pressure and the pipe diameter is c. The pressure (Δp) must be just enough to overcome the hydrostatic pressure of the fluid at depth (H_{ft}) to form a bubble. Higher flows will just produce more bubbles. Determining the initial bubble formation, point will be a start to understanding the magnitude of the problem.

When the volume in the bubble (V) increases to the point that the force trying to hold the bubble to the pipe $(F_{tension})$ and the buoyancy force $(F_{buoyancy})$ which is trying to pull the bubble away are equal, just a little more buoyancy will cause it to float to the surface. The buoyancy force is equal to the weight of the amount of liquid that the formation of bubble volume displaces and is sometimes called Archimedes' principle. It's the reason heavy steel ships float.

The flow (q) out of the pipe will be considered to be the amount required to just achieve (Δp) at the submerged depth (H_{ft}).

The problem is certainly much more complex than this and Figure 4.10.2 shows the actual case of the necking off of the bubble before it tears loose and seals itself. This means it overcomes the tension. This is evident when blowing soap bubbles through a small hoop. It stretches out, meaning it elongates like a large pouch, seals, and floats away.

This is the necking down model and is a complex one and requires knowledge of the membranes properties and other phenomena. All that is desired in this analysis is a rough approximation of the frequency the bubbles are being released at with a constant flow into the pipe (q) and the magnitude of the excitation force as it breaks loose, which is also the buoyancy force $(F_{buoyancy})$. Figure 4.10.3 is one of many simple stop action bubble experiments with a straw used to determine the constant (φ).

The idea of how a steady-state flow can turn into a cyclic flow will also be understood from such an analysis. After all when you blow into a straw in a glass of water, it bubbles. The harder you blow, the more violent the bubbles are and the straw

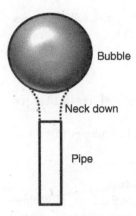

Figure 4.10.2 Bubble necking down.

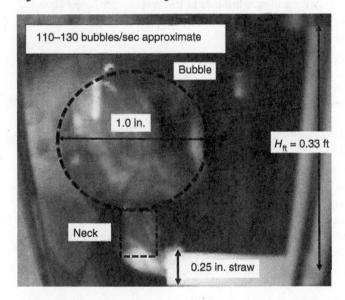

110–130 bubbles/sec approximate

Bubble

1.0 in.

$H_{ft} = 0.33$ ft

Neck

0.25 in. straw

Figure 4.10.3 Experiment with straw.

shakes a little. The deeper the straw goes into the water, the harder it is to blow that first bubble. It all looks like it should be simple to analyze, but it isn't.

The buoyancy force ($F_{buoyancy}$) at a depth (H_{ft}) and the flow rate (q) out of the pipe can be approximated from Bernoulli's equation discussed in Section 12.11;

$$\Delta p = (H_{ft})^*14.7/34 \text{ lb/in.}^2$$

$$q = 2{,}500^*c^2{^*}\Delta p^{1/2} \text{ in.}^3/s$$

$$F_{buoyancy} = ((\pi/6)^*d^3)^*\rho$$

where d is the diameter of the bubble that needs to be determined.

The tension force (F_{tension}) trying to hold the bubble to the pipe will be *approximated* by a simple experiment of blowing bubbles in water through a straw as shown in Figure 4.10.3.

$$F_{\text{tension}} = \varphi*(\pi/4)*c^2$$

where φ is a constant determined by the simple experiment and represents a tensile stress.

Equating (F_{buoyancy}) to (F_{tension}) and solving for the bubble diameter (d):

$$d = [3*\varphi*c^2/(2*\rho)]^{1/3} \text{ in.}$$

The value of φ is determined by the size of a bubble and is $\varphi = 0.4$.

This holds for the small-scale experiment of $c = 0.25$ in. and 0.75 in. pipe diameter at a depth (H_{ft}) of 1/3 ft. It is unknown how valid it is scaled up to $c = 4$ in. at a depth of 2 ft. Most likely the full-size bubbles in the vessel were larger than the pipe diameter of 4 in.

The time and frequency of bubbling is:

$$t = \text{volume of a bubble}/q = ((\pi/6)*d^3)/q \text{ s}$$

The frequency is $f = 1/(t)$ bubbles/s.

The force can be considered as the buoyancy force (F_{buoyancy}) being released from the pipe. It will react on the pipe as it pushes away. Higher flow rates will result in higher frequencies. The results on the experiments and examples are shown in Table 4.10.1.

Actual failures of spargers due to vibration seem to validate the direction of the analysis and that is that gas bubbling into a liquid can cause cyclic excitation forces. These are strong enough to fail stab-in tube bundles excited by a steam as shown in Figure 4.10.4. It is also included in Table 4.10.1. The bubbles are assumed to exit the holes of the spargers at the same time since the pressure acting on each of the holes is the same. Summing the individual forces results in a total force which is resolved as a moment on the tube bundle for a stress analysis.

Table 4.10.1 Flow out pipe and excitation frequency.

Description	Pipe diameter (in.)	Bubble diameter (in.)	Frequency (bubbles/s)	F_{bubble}
Tank $H_{\text{ft}} = 2$ ft	4	6.5	270	5 lb/bubble
Sparger $H_{\text{ft}} = 5$ ft 500 holes	0.25	1	420	0.02 lb/bubble or 10 lb for 500 bubbles at same time
Straw $H_{\text{ft}} = 0.33$ ft	0.25	1	110	0.02 lb/bubble
Pipe $H_{\text{ft}} = 0.33$ ft	0.75	2	110	0.2 lb/bubble

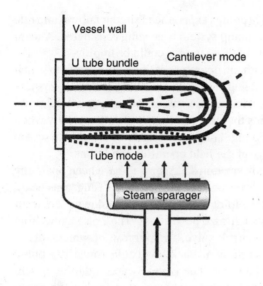

Figure 4.10.4 Sparger exciting a tube bundle.

4.10.1 Summary

For the tank pipe vibration, even though the bubble frequency was far from the pipe lateral natural frequency, the forcing magnitude could be one of the factors causing the rattling-type vibration. It's unlikely that reducing the flow would have much of an effect. The solution was a design modification that limited the pipe motion.

The sparger problem was solved when a process engineer said the steam flow could be greatly reduced, which eliminated the violent bubble release because of the high flow.

Blowing through a straw into a glass of water shows the increase in frequency and increased violence as the flow is increased.

The bubble formation model was not accurate but did help in understanding what was occurring with forces and frequencies. This is the case in many practical engineering solutions.

4.11 Failure of a Cooling Tower Pump Due to Water Hammer

A large cooling tower pump, which was one of several, kept having failures with its seals, base plate hold down bolts, cracking of the foundation, gasket leaks, and check valve failures. These problems were addressed by just repairing them with no investigations on why they were occurring. The cost of these repairs over the years was more than \$100,000. All five pumps experienced similar failures and due to the high maintenance cost, a new pump system was being considered. A team was assembled to see if the reliability of these pumps could be improved.

One of the first steps was to talk with the operators of these pumps and to watch the start-up and shutdown of them. Being at the site, watching something operate is always beneficial as it provides firsthand observations of any procedural problems. It's difficult to make statements that the pump is operating correctly when product is spraying all over one's clothing. This happens! All too often analyses are done in the office with no knowledge of the field operating conditions.

In the case of one of the cooling tower pumps, it seems to have been operating correctly. One day while observing a routine shutdown, a loud "Bang" was heard that shook the ground and rattled the piping. Uncontrolled shutdowns can occur due to motor trip-outs for reasons such as a power loss. The noise came from the location of the large discharge check valve, the purpose of which was to stop backflow into the pump on shutdown. Backflow could rotate the pump in the opposite direction than designed for. The consequence could be loosening the impeller, damaging seals, or bearing problems by spinning the pump in reverse.

This violent slamming shut of the check valve is usually associated with a dynamic condition called "water hammer." Most of us have heard it in our homes when we suddenly shut off water at a sink with a lever-type valve. There is a "thump" and the pipes may rattle. When the water valve is closed slowly, this usually doesn't occur.

The water hammer phenomena can be so damaging in industrial installations that will be explained in some detail. Consider the pump system of Figure 4.11.1.

The force build-up at the check valve due to the head is:

$$F + F_o = F/2$$

$$F = (W/g)*a \quad \text{where } a = (V_f - V_i)/t = -V_i/t$$

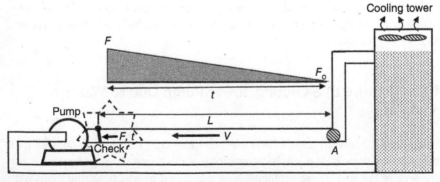

Figure 4.11.1 Cooling tower pump circulation system.

$$F = (W/g)*(V_i/t)$$

$$(1/2)*F/A = (W/(g*A)*(V_i/t) \text{ but } W = \rho*L*A$$

After some manipulation:

$$\Delta p = 0.028*L_{ft}*V_{ft/s}/t_s \ \text{lb/in.}^2$$

The force on the valve is simply the pressure times the pipe cross-sectional area that simplifies to:

$$F = (1/2)*(0.028*V*L*A)/t \ \text{lb}_f$$

The $(1/2)$ is because of the average linear build-up of the pressure.

This is valid only when $t > 2*L_{ft}/4{,}700 \ \text{ft/s}$.

The value 4,700 ft/s is the speed of sound in water or the time it takes for the shock wave to rebound.

In a check valve on the discharge piping, the flow velocity (V) is not the normal forward flow as in a water faucet but is the backflow as the flow is reversed and passing through the not as yet closed check valve. The longer it takes to close, the higher the flow through it as the valve is closing. Notice that now the longer the valve is open, the higher the force, which is the opposite of shutting off a faucet.

When the pump stops, the flow toward the cooling tower ceases. The check valve is still open and is starting to close so the flow toward the pump is due to the pressure head H_{ft} pushing the water toward the pump as shown in Figure 4.11.2.

The backflow velocity can be calculated from a simplified form of Bernoulli's equation [1]:

$$V_{backward} = (2g*H_{ft})^{1/2} \ \text{ft/s, where } g = 32.2 \ \text{ft/s}^2$$

This is the maximum backward flow when the check is wide open and full flow has developed. However, this velocity can't be developed in such a short time [2]. Assume only one-fourth of this velocity is reached or $V_{backward}/4$.

This now is similar to the flow equations developed.

$$F = (1/2)*(0.028*V_{backward}*L*A)/t \ \text{lb}_f$$

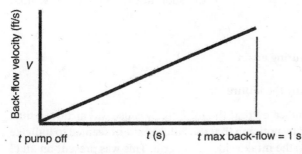

Figure 4.11.2   Backflow when pump stops.

With $L = 100$ ft, $A = 115$ in.^{2}, $t = 1$ seconds, $H = 15$ ft, $V_{backward} = 7.5$ ft/s, $F = 1,250$ lb.

4.11.1 Summary

Closure of check valves can result in high-impact loads on the pump and piping system. The design of the check valve greatly influences this force. Ball and swing check valves develop the highest forces. After reviewing the system, a check valve manufacturer suggested a better valve that was installed and solved the problem. It was much more cost-effective than new pumps.

Here is a case where the problem source was identified and the solution was provided by a manufacturer's valve specialist. There was no need for an exact analysis to the problem; however, the analysis provided direction to the team.

References

1 Sofronas, A. (2006). *Analytical Troubleshooting of Process Machinery and Pressure Vessels*, 143. Wiley.
2 How to prevent circulating water flow reversal. *Power Magazine*, 30 May 2017.

4.12 Braking Resistor Burn-Out on a Locomotive

This is a case history that utilizes some of the equations used in this book. One of the braking resistor grids on a locomotive, which is used to dissipate power, had been burning out. Diesel–electric locomotives use their traction motors as generators in dynamic braking to slow the locomotive down. These units contain a length of high-temperature bent metal ribbon similar to Figure 4.12.1.

The power (Q) that needs to be dissipated is passed through the high resistance ribbon. This is similar to what occurs in a home bread toaster's resistance wires. Air is forced over the ribbon (V) by a fan to remove the heat. Periodically, the ribbons melted from too high a temperature and the question was what could cause this? Some possibilities are:

1. airflow reduction
2. poor design or manufacturing error
3. excessive power applied
4. damage to the grid causing the failure

A review of the failures indicated that the most likely cause was airflow reduction. This was substantiated by noticing that the failure pattern seemed to indicate that something had stuck to the intake side of the grid. This was present on all of the failures and seemed to cover 25% of the grid surface. This would have reduced

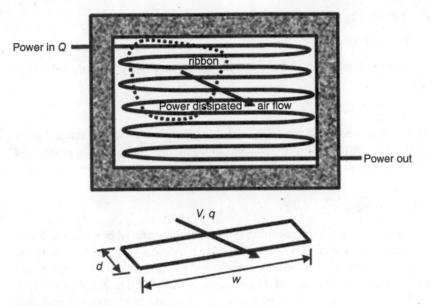

Figure 4.12.1 Ribbon braking resistor.

the airflow due to the obstruction. The following analysis was performed to see if this was a possible cause. A phantom-type failure would have occurred if debris hadn't been observed.

Consider that the ribbon is simplified to consist of a series of (n) flat plates each with a width (w) and a depth (d). Over each is flowing forced air at a velocity (V). As it passes over each plate, it removes (q) amount of heat by forced convection. The challenge is to determine the temperature of the ribbon (i.e. flat plate) with and without the area blocked off to see if it will melt.

The grid is designed to dissipate Q_{in} BTU/h where (Q_{in}) = 2.546*horsepower dissipated.

Air flowing over the plates removes the majority of the heat and radiation dissipation is not considered.

Consider that an individual plate will dissipate $q = Q_{in}/n$.

Air is flowing over both sides of the plate so the heat removed is:

$$T_{plate} = (q)/(h*A) + T_{ambient}$$
$$A = 2*d*w$$

For forced convection, an approximation is:

$$h = 1 + 0.225*V \text{ BTU}/(\text{h ft}^2\ ^\circ\text{F})$$

Consider $Q = 382{,}000$ BTU/h (i.e. 150 HP), $n = 30$, $d = 0.25$ ft, $w = 3$ ft, $V = 40$ ft/s, $T_{ambient} = 110\,^\circ$F.

With full velocity over both sides of the ribbon and no blockage:

$$T_{plate} = 960\,°F$$

Now assume blockage causes the velocity over the ribbon to be only 25% of the velocity:

$$T_{plate} \approx 2,700\,°F$$

The ribbon is made of Inconel with a melting temperature in this range. Under full braking conditions, with no blockage, at night, the ribbon was an orange-red color that is around 1,000–1,500 °F.

4.12.1 Summary

It appears that severe blockage could cause the braking grid ribbon to melt. It was later noticed that paper and plastic debris were periodically sucked in front of the grid restricting airflow. Sometimes, this caused the grid to burn out before the debris either burned off or fell off. This is shown as the dotted line in Figure 4.12.1. The solution was to design a standoff corrugated screen in front of the grid inlet to keep debris off of the surface. This allowed air to flow around the debris and through the grid.

4.13 Will a Small Ice Air Conditioner Work?

Engineers are always coming up with ideas. I have a book full of my own. Many are because a situation has occurred that they think they could help with. Engineers have the advantage that they really don't have to build anything to see if an idea is worthwhile, they can do it with calculations. Here's an example of this.

Consider a hot day where people are suffering in their homes from a 80+ °F. room temperature. They can't afford to pay for a true air conditioning system, repairs or the electricity they use. A fan alone doesn't provide the cooling needed. They have a refrigerator with a freezer section and have to pay for that electricity. What could someone buy that is inexpensive and would cool them down more than a fan?

The goal isn't to cool a room. All that is desired is to be cooled more than a small fan would do. A fan cools you down by evaporating the sweat off of you as described in Section 8.5 and doesn't cool the air. Wet your hand and blow air on it and you'll see this effect. This is what these coolers do which are really only humidifiers that blow air over a wet pad. They are also known as swamp coolers. Who needs a humidifier in Florida?

Ice chest cool air systems are available and those that blow air through the ice are best. They are bulky but can provide very cold air for several hours.

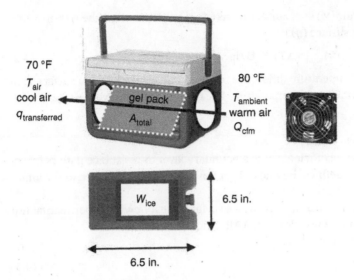

70 °F
T_{air}
cool air
$q_{transferred}$

gel pack
A_{total}

80 °F
$T_{ambient}$
warm air
Q_{cfm}

W_{ice}

6.5 in.

6.5 in.

Figure 4.13.1 Gel ice pack cooler.

This design analyzes the possibility of a very small, low-cost ice cooler that is light and easily carried like a small fan.

A small gel ice pack, available in stores and used in picnic coolers, is frozen in the refrigerator freezer. A small personal cooling system based on this will be analyzed. The gel pack is envisioned as fitting into a picnic-type cooler with a small 115- or 12-V fan blowing ambient air on both sides. The warm air would be blown over the gel ice pack surfaces and would cool the air as shown in Figure 4.13.1, without humidity.

The idea sounds simple enough but there are questions. How long will the gel pack cool before having to be replaced with another from the freezer? What is the temperature of the cooled air? How far from the discharge will it cool someone? Some calculations can help quantify this.

A block of ice left in a warm space melts as it changes from a solid to a liquid. As it does it absorbs heat energy. Removing heat is what cools the air.

The latent heat of fusion is the heat energy needed to change ice to water. It is the amount of energy available to cool something down. Once it's used up, no more cooling is possible. From thermodynamic property tables, the amount of energy absorbed is 144 BTU/lb of the 32 °F ice used.

$$\Delta E = 144 * W_{ice} \text{ BTU} \tag{4.1}$$

where W_{ice} is the weight of the ice in pounds.

The first question is how long will this amount of energy last?, meaning how long before the ice all turns to cold liquid water?

For the flow of air (V) over surface areas (A_{total}), the amount of heat transferred from the gel pack surface (q) is:

$$q_{transferred} = h*A_{total}*(\Delta T) \text{ BTU/h} \tag{4.2}$$

where ΔT is the temperature difference between one gel pack's total surface area (A_{total}) and the ambient air temperature. The gel pack surface averages 42 °F with an airflow over it.

$$\Delta T = (T_{amb} - 42)\,°F$$

The heat transfer coefficient (h) is a boundary layer or resistance path between the ice and air that defines the rate of heat flow from the block of ice to the ambient air.

Since h depends on airflow over the ice surface, the following experimental test shown in Figure 4.13.2 was done on 0.7 lb of ice.

The equation of this curve is:

$$h = 0.75*V + 6 \tag{4.3}$$

The end of the test was when the ice had melted. The cold water left was still at the ice temperature so all of the energy wasn't used. The life to ambient water temperature might have been 50% longer.

Using Eqs. (4.1)–(4.3), the time required to use up ΔE can be determined.

Assume that a gel pack (6.5 in. × 6.5 in. × 2 in.) is used and air is flowing over both sides of the ice pack with the following values:

$$Q_{cfm} = 70 \text{ ft}^3/\text{min}, \ A_{total} = 0.58 \text{ ft}^2, \ W_{ice} = 4 \text{ lb}$$

$$q_{transferred} = 301 \text{ BTU/h}$$

$$\Delta E = 144*W_{ice} = 576 \text{ BTU}$$

To transfer all the heat available from the one gel pack at 42 °F and bring it to T_{amb}:

$t = (\Delta E/q_{transferred}) = 1.9$ hours and could be 50% more if the melt is also considered.

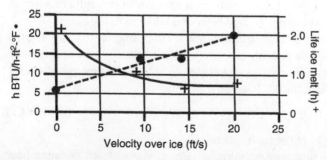

Figure 4.13.2 Forced convection heat transfer coefficient versus velocity.

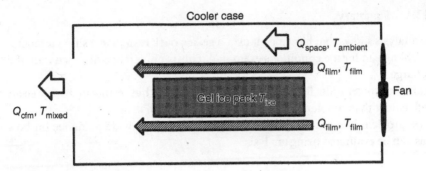

Figure 4.13.3 MIxed flow.

Some idea of the temperature in the airflow at the discharge (T_{air}) can be estimated by using mixing flow theory.

The problem now becomes a little more difficult because the amount of flow (Q_{cfm}) removing the heat is a mixed flow condition. The velocity over the ice surface is a boundary-layer film that is only a small percentage of the total flow but is at the ice temperature. Figure 4.13.3 illustrates this.

From laminar flow theory [1], the total cold film zones represent 5% of the flow and the space away from the flow represents 95%. Weight flow is usually used but since the densities aren't much different and humidity isn't considered, volume flows are used.

From flow mixing used in air conditioning ducting systems based on percent flows [2];

With two surfaces exposed:

$$Q_{cfm} = 0.05*Q_{film} + 0.95*Q_{space}$$

$$T_{mixed} = 0.05*T_{film} + 0.95*T_{ambient}$$

$$T_{mixed} = 0.05*42 + 0.95*81 = 79\,°F \text{ downstream temperature}$$

Ice life = 1.9 hours.

Consider a large commercial insulated forced air ice cooler with a 50% split. The T_{mixed} would now be:

$$T_{mixed} = 0.5*42 + 0.5*81 = 61.5\,°F$$

The air velocity (V ft/s) of this mixed temperature out of the discharge is:

$$V = (Q_{cfm}/(0.785*D^2_{disch}))/60 \tag{4.4}$$

This velocity should be greater than 20 ft/s so it can be felt some distance away. With $D_{disch} = 0.39$ ft, the discharge velocity (V) = 10 ft/s.

In Section 4.14, an experimental model is built and data gathered to validate the analysis.

4.13.1 Summary

The analysis shows that blowing air over a gel ice pack results in a slight temperature change under the conditions examined. This is not felt at a distance from the discharge.

Large ice cooler units flow cold air in the cold space that results in discharge air much colder than ambient air. This is shown in Section 4.14.

The problem is more complex than this but it does provide guidance on how ideas can be evaluated using analysis.

References

1 Holman, J.P. (1963). *Heat Transfer*, 102. McGraw-Hill, Inc.
2 Richardson. (2019). Four steps to determine the correct mixed air temperature. *ACHR News*, 18 Nov 2019.

4.14 Prototype of Smallest Air Ice Cooler

Very rarely have I performed an analysis that isn't backed up with some sort of verification data. There are just too many unknowns and I enjoy building models.

The analysis of Section 4.13 shows that a small gel pack ice cooler would produce slightly cooler air. The real interest in building the prototype was to ensure that the analysis procedure was reasonable.

Figure 4.14.1 shows the unit built and tested. The test unit was made from an available gel ice pack and a small computer cooling fan in a cardboard case with

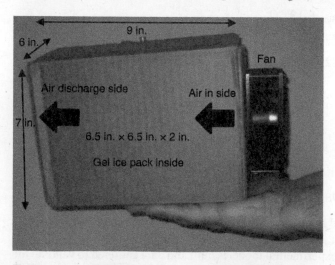

Figure 4.14.1 Prototype small gel ice pack cooler.

Table 4.14.1 Calculated and actual results for prototype.

Ambient air 81 °F	Q_{fan} (cfm)	Exit velocity (ft/s)	$q_{transferred}$ (BTU/h)	Effective life (hours)	T_{mixed} (°F)	ΔT (°F)	Temperature felt at distance (ft)
Test results 6.5*6.5*2 in.^{3} gel pack in box two sides exposed to flow $W = 4$ lb	70	10	Not measured	2½ h pack liquid cold	78.0	3.0	Not felt
Calculate two sides exposed $W = 4$ lb	70	10	301	1.9	79.0	2.0	Not calculated
7-gal bucket, 30-lb ice frozen in 4 plastic containers, flow around containers	175	10 calculated	1,500 calculated	2.9 calculated	62.0 calculated	19.5 calculated	Felt about 2 ft away

no insulation. The logic is that if it works with no insulation it will only perform better insulated. Not a very pretty design but good enough for a quick test.

The cooling fan produced 70 cfm through the small box. Test data was conducted on the unit to measure the cooling ability and to compare it with calculated data. This is shown in Table 4.14.1.

This very small ice cooler had a 3-in. diameter discharge nozzle to increase the velocity of the cooled air to 10 ft/s. Two sides of the gel pack were exposed. The slightly cooler air was not felt away from the discharge.

The last row in Table 4.14.1 is for a product being sold. This smaller ice cooler cools the space before being discharged. Customer reviews were not good. Some said they had to sit right next to it to feel any cool air and others just sent the unit back because of the poor performance. The calculations indicate that the discharge was cold but the fan was too weak. A higher flow fan would have shortened the ice life. The manufacturer states a 7-hours life but calculations indicate in the 2- to 4-hours range.

With larger units and higher flows that are designed to go between ice cubes instead of off of the top, the discharge temperature can get as low as 42 °F. These large coolers with high-volume fans can last 2–3 hours on a hot day and seem to work well.

4.14.1 Summary

The small design as built isn't useful as it just doesn't have enough ice and cooling volume. The mathematical model of Section 4.13 explains this. The analysis and test model are in reasonable agreement and explain how these units function.

The reader will appreciate that before the analysis, at the start of Section 4.13, little was known about the workings of such a small ice cooler and if it would or wouldn't perform useful cooling. After the analysis, it was obvious that it wouldn't be useful as designed and why it wouldn't. All the prototype model did was to confirm this.

5

Sports Examples

Curiosity Is An Engineers Gift

5.1 Why Does a Baseball Curve?

The game of baseball is an intriguing sport that I was never very good at. That doesn't mean I don't have a great respect for players that do it well. I'm especially in awe of the pitcher and catcher. The catcher is described in Section 5.4.

The pitcher can do some wondrous things with the baseball, besides just throwing it fast and accurate. The ball can be made to turn in flight by the pitcher. The flight path of the ball will normally drop as it nears the plate due to the vertical force of gravity on the ball and the slight loss of speed due to air friction. However, the ball can also be made to curve sideways, downward, or even rise by the pitcher's throwing technique. The curve can deviate sidewise from a straight line by up to 18 in. at home plate. How is this possible?

Figure 5.1.1 illustrates the velocity (V) of the baseball, as it is thrown. By configuring the fingers on the ball during the pitch and twists of the wrist, the pitcher can impart a spin to the ball of up to 2,200 rpm. As is shown, because of the rotation, called backspin in this case, the velocity on one side of the ball is higher than the other side.

The lift or push on the baseball is provided by the lift equation where the subscript n is the top or bottom of the baseball. This says the flow of air over a surface produces a lift (L_n) on the surface.

$$L_n = (\tfrac{1}{2})^*C_L{}^*(\rho/g)^*A^*V_n^2$$

Unique Methods for Analyzing Failures and Catastrophic Events: A Practical Guide for Engineers, First Edition. Anthony Sofronas.

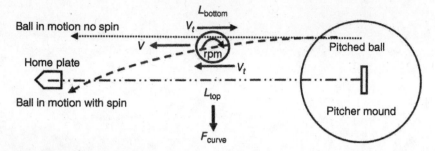

Figure 5.1.1 Model for curveball.

where C_L is the lift coefficient of a baseball which is affected by spin and is typically 0.14, ρ is the density of air or $5 * 10^{-4}$ lb$_f$/in.^{3}, $g = 386$ in./s^2, A in.^{2} is the projected area of a baseball, V_n is V_{top} or V_{bottom}. The variable V is the velocity of the air over the baseball without spin. The following is the air velocity on the baseball:

$$V_{top} = V + V_t$$

$$V_{bottom} = V - V_t$$

The tangential velocity due to the spin of the baseball is:

$$V_t = \pi*d*rpm/60$$

With $d = 3$ in., rpm $= 2,200$, $V = 1,584$ in./s (90 mph).

$$V_{top} = 1,700 \text{ in./s}, V_{bottom} = 1,470 \text{ in./s}$$

From Table 5.1.1, it is obvious that the pitcher can control the ball with both the speed of the pitch and also the spin of the baseball since it changes the force on the baseball. Where the ball curves depend on factors such as how much and how long the air is disturbed by the stitches of the baseball and a lot of other things the pitcher does, based on experience.

Table 5.1.1 Effect of spin and speed on curve ball.

Condition	Velocity pitch [mph (in./s]	Spin (rpm)	$L_{top} - L_{bottom}$ (lb$_f$)
Fastball with spin	90 (1,584)	2,200	0.47
Fastball low spin	90 (1,584)	1,100	0.23
Slow ball with spin	70 (1,232)	2,200	0.36
Slow ball low spin	70 (1,232)	1,100	0.18

5.1.1 Summary

By controlling the spin on the baseball and speed of the pitch using various finger, arm, and wrist-twisting techniques, an experienced pitcher can have great control over where the ball is directed to the catcher's glove.

5.2 How Far Does a Baseball Go When Hit with Drag?

We see the swing and hear the crack of the bat and for a few seconds, we wonder how far the ball will go.

To determine the ball trajectory with drag on the baseball, meaning with air resistance but no spin, the model of Figure 5.2.1 will be used.

We will assume that a radar camera recorded the ball leaving the bat at 110 mi/h or at an initial velocity $V_i = 161$ ft/s.

Figure 5.2.2 is the forces on the baseball.

The force component is due to the baseball accelerating in the vertical direction:

$$F_{av} = (W/g)*a_v$$

$a_v = V_{vi}^2/(2*s)$ where $V_{vi} = V_i*\sin(\theta)$.

The force component is due to the vertical drag on the baseball:

$$F_{dv} = \left[(1/2)*C_d*(\rho/g)*A*V_{vavg}^2\right] \text{ where } V_{vavg} = V_{vi}/2$$

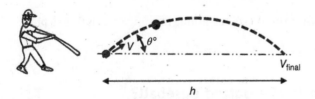

Figure 5.2.1 Baseball trajectory without drag.

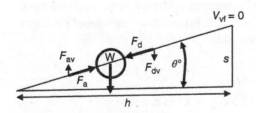

Figure 5.2.2 Forces on baseball.

Using a force balance on the baseball until there is no more vertical lift but it still has horizontal velocity to come back down:

$$F_{av} - F_{dv} - W = 0$$

These equations can be solved for s and as a rough estimate $h = 2*s$.

For a baseball $V = 161$ ft/s (110 mph), $g = 32.2$ ft/s^2, $\theta° = 45$, $C_d = 0.3$, $A = 0.046$ ft^2, $\rho = 0.078$ lb/ft^3, $W = 0.33$ lb:

Range $h = 402$ ft with $C_d = 0$, $h = 346$ ft with $C_d = 0.3$.

From the equations, 45° is used because it produces the greatest distance.

Typically, a home run of 350 ft is possible and the longest ever verified in the air with no rolling was in the 500-ft range. So drag reduces the flight by about 14%. Drag depends on many factors such as the ball roughness, spin, and deformation.

This is why it's so hard to get a long hit. Everything has to be just right. Where you hit the ball, with how much force, where on the bat it hits, the angle of your swing, the condition of the ball, the spin on the ball, the wind, and there are many other factors such as luck.

I say this to make myself feel good. In my youth, I wasn't a promising baseball player.

5.2.1 Summary

How far a baseball is hit depends on many factors. Over 500 ft but less than 600 ft have been done and is possible. A distance of 350 ft is a reasonable home run distance.

These types of equations were used in determining the range debris fly in a pneumatic explosion.

5.3 What Is the Force of a Batted Baseball?

I always wondered this even before I was an engineer. Assume a baseball is thrown at 90 mph (132 fps) and the bat tip is also moving at 90 mph in the opposite direction. A standard baseball weighs 5.125 oz (0.32 lb) and from high-speed photography, the contact time is about 0.0007 seconds.

Using Newton's Second Law:

$$F_{avg} = (W/g)*(V_{ball} - (-V_{bat}))/\Delta t$$
$$= (0.32/32.2)(132 + 132)/(0.0007) = 3{,}750 \text{ lb force}$$

This type of analysis was used on analyzing the impact of product on a mixer blade that had broken its attachment bolts. A simple analysis identified the cause and the redesign that was necessary.

Sometimes, one wonders why this force isn't felt on a batter's hand. It can be shown that there is a point on the bat called the "center of percussion" where there is no reaction on the batter's hand and the mass seems centered here. Think of a hammer hitting a nail. You don't feel that on your hand because the hammerhead is the center of mass and most of the work is done by the hammerhead weight and its inertia.

5.3.1 Summary

The force on the bat can be 3,750 lb and the hand doesn't feel it because it's absorbed by the mass of the bat and energy into the ball.

5.4 Why Doesn't a Baseball Catcher's Arm Break with a 100-mph Fastball?

When watching baseball, it's always amazing to see a catcher reacting to each pitch. It is said that it stings and the hand hurts after the game and has to be iced down. There have been cases where an arm was broken while catching the ball.

The following model shown in Figure 5.4.1 will be used to examine the force (F) on the catcher's glove.

In this model, W_1 is the weight of the baseball, W_2 is the weight of the catcher's glove and arm, V is the velocity of the baseball when it reaches the glove, k is the spring constant of the glove and body flexibility established by a test, and δ is the amount of motion when W_1 stops in the glove.

Figure 5.4.2 is the actual model analyzed.

Drawing by Allyson Sofronas

Figure 5.4.1 Model of a catcher reacting to a fastball.

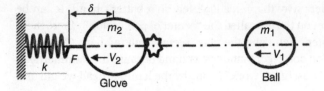

Figure 5.4.2 Lumped mass model.

Equate the momentums assuming the masses stick together and don't bounce:

$$m_1 V_1 = (m_1 + m_2)^* V_2 \text{ then } V_2 = [m_1/(m_1 + m_2)]^* V_1$$

The kinetic energy becomes:

$$KE_2 = (\tfrac{1}{2})^* m^* V^2 = (\tfrac{1}{2})^*(m_1 + m_2)^*([m_1/(m_1 + m_2)]^* V_1)^2$$

The potential energy PE_s of the spring (k) is:

$$PE_s = (\tfrac{1}{2})^* k^* \delta^2$$

Equating KE_2 to PE_s as the spring absorbs KE_2 and solving for the displacement of (m_2):

$$\delta_{spring} = [m_1/(m_1 + m_2)]^* V_1^*[(1/k)^*(m_1 + m_2)]^{1/2}$$
$$\text{where } m_1 = W_1/g \text{ and } m_2 = W_2/g$$

The catcher can also react by pulling his glove back (δ_{react}):

$$\delta_{total} = \delta_{spring} - \delta_{react}$$

The force in the catcher's glove is therefore the reaction:

$$F = k^* \delta_{total}$$

5.4.1 Some Data

A catcher's glove can weigh up to 2 lb and the forearm portion can weigh 3 lb based on ergonomic studies, so $W_2 \approx 5$ lb. The baseball W_1 weighs 0.33 lb and V_1 will be assumed to be 100 mph (147 ft/s). The value for k was the result of a force deflection test in a catcher's position. A force was put on the glove and the distance deflected in the web measured. This resulted in $k =$ Force measured/amount pushed back = 360 lb/1 ft = 360 lb/ft. In one case, the catcher moved the glove back 1 in. and the other catcher doesn't.

The calculations' results are shown in Table 5.4.1.

Table 5.4.1 shows that the catcher has some control of the force his hand receives. By being more relaxed the force is lower. How the catcher puts the ball in the glove and the catcher's reaction also makes a big difference.

Table 5.4.1 Force on catcher's mitt.

Case 100-mph pitch	k (lb/ft)	Force in glove (lb)
Catcher moves mitt 1 in.	360	40
Catcher doesn't move mitt	360	70

5.4.2 Summary

A 100-mph fastball will impact the catcher's glove with 70 lb of force. The catcher has additional control on reducing this force by the way the ball is caught in the glove.

5.5 Dynamics of a Billiard Ball

When playing billiards, it has always amazed me at how spin on the ball could have such an effect. How does spin control how far the ball will go?

Consider Figure 5.5.1 where the ball is impacted as shown. It will impart a spin on the ball (ω_i) and also move the ball at a velocity (V_i). The spin will react with the mass of the ball (m) and the friction on the surface (μ) will try to slow the spin. With pure rolling, there is no frictional effect, like when the ball just starts to roll at $t = 0$. However, as the ball is accelerating to V_f, friction slows the spin until there is no more spin at $t = t$. At this point, the velocity is pure rolling again. So the spin has slowed down the velocity. What is happening mathematically? The conservation of angular momentum can be used to explain this.

The ball or sphere angular moment of inertia is $I = (2/5)*m*R^2$.

When the cue hits the ball, it has a linear and angular momentum at the start:

$$V_i = \omega_i*2*R \text{ and } V_f = \omega_f*R$$
$$L_{start} = m*V_i*R + 2/5\, m*R^2*(V_i/(2*R))$$
$$L_{end} = m*V_f*R + 2/5\, m*R^2*(V_f/(*R))$$

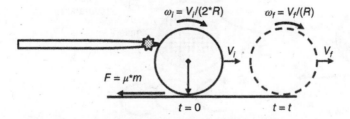

Figure 5.5.1 Impact of cue on ball model.

By conservation of momentum $L_{start} = L_{end}$ and using some mathematical gymnastics:

$$V_f = (6/7)*V_i$$

Since rotation is no longer causing friction, only the mass of the ball on the table causes friction so $a = \mu*g$.

$$V_f = V_i - a*t$$

$$(6/7)*V_i = V_i - \mu*g*t$$

The time for the rotating friction to become zero is:

$$t = (1/(7*\mu*g)*V_i)$$

5.5.1 Summary

By putting spin on a billiard ball, you can shorten the distance it will go as it affects the retarding friction and essentially puts friction drag on the ball. When the cue is off center, it will impart spin in another direction and will cause the ball to veer in that direction.

5.6 How Far Can a Golf Ball Go?

The physics of hitting the golf ball is similar to that of batting a baseball. Instead of stitches like on the baseball, the golf ball has dimples. Dimples change the flow of air in flight from laminar flow to turbulent flow. The turbulent zone caused by the dimple turbulators reduces the drag. Laminar and turbulent flows are illustrated in Figure 5.6.1.

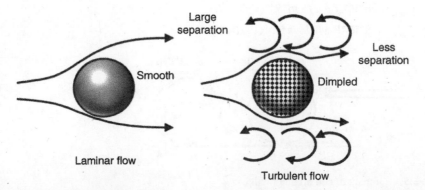

Figure 5.6.1 Flow due to dimples.

The drag coefficient for a smooth sphere is about $C_d = 0.5$ and for a dimpled golf ball $C_d = 0.2$. This dimple design can vary from 0.12 to 0.2 depending on the type and condition of the ball.

Using the same drag and distance equations from Section 5.2, the distance of a golf ball with no roll can be estimated. The ball weighs about 0.11 lb$_f$ and has a cross-sectional area of 2.2 in.[2] with a ball diameter of 1.68 in. Statistics say the average golfer's ball velocity of a drive is $V_i = 200$ ft/s. Assume it has a departure angle of 45°.

Distance $C_d = 0.4, s = 465$ ft

Distance $C_d = 0.15, s = 550$ ft

5.6.1 Summary

For an average golfer, the dimples on a golf ball add about 18% to the distance of the drive over that of a smooth ball. This is due to an increase in turbulence that causes a reduction in drag.

These types of equations were used in determining the range debris would fly in an explosion.

5.7 What Causes an Ice Skater to Spin so Fast?

We have all watched skating competitions where the skater starts to rotate in place and then pull their arms in and spin real fast. What's happening?

Figure 5.7.1 illustrates this.

Notice what is happening. The skater has her arms stretched out (r_o) and starts the spin (ω_o) of her mass (m_o) around the vertical axis. She then pulls her arms in r_{in} and the spin (ω_{in}) of her mass (m_{in}) around the vertical axis speeds up.

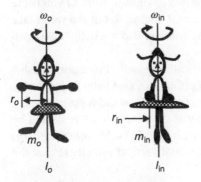

Figure 5.7.1   Ice skater spinning.

This can be explained by using the conservation of angular momentum where L is angular momentum or $L_{armsin} = L_{armsout}$.

Angular momentum is defined as:

$$L = I*\omega$$

where I is the angular moment of inertia of the rotating figure skater with her arms either in (I_{in}) or out (I_o) and ω is the angular speed of rotation with arms in (ω_{in}) or out (ω_o).

$$I = m*r^2$$

For simplicity, let's say that $m_o = m_{in}$ and $r_o = 24$ in. and $r_{in} = 8$ in.

This means $I_{out}/I_{in} = 9$ and the skater's moment of inertia is 9 times more when arms are out than when arms are in.

From the conservation of angular momentum,

$$\omega_{in}*I_{in} = \omega_{out}*I_{out}$$

$$\omega_{in} = \omega_{out}*I_{out}/I_{in}$$

The spin of the skater with arms in (ω_{in}) will therefore be 9 times faster than with arms out (ω_{out}).

5.7.1 Summary

An ice skater's spin is increased when the arms are folded in to conserve angular momentum.

5.8 Why Don't High Divers Get Injured?

Diving into smooth water head first with your arms straight in front of you is the normal way to dive and results in little discomfort. Landing flat in the water face first with a loud smacking sound is another way and is called the belly flop which is far less graceful and more painful.

A record belly flopper is Darren Taylor, who calls himself "Professor Splash," just broke the Guinness World Record by diving from 36 ft and belly flopping into a 7-ft diameter kiddy pool filled with about 1 ft of water. How did he survive?

Consider Figure 5.8.1 where the deceleration of the diver's body is slowed by pushing the water out of the pool. Diving from a height (h) with only gravity (g) pulling him toward Earth and no air drag, the initial vertical velocity (V_i) at the water surface is:

$$V_i = (2*g*h)^{1/2}$$

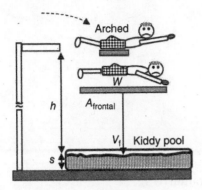

Figure 5.8.1 Belly flop into pool.

When the body hits the water, it will displace a volume of water and be cushioned until it comes to a stop. It has to accelerate the mass (W_w/g) of water out of the way.

The deceleration from the surface of the water to the distance (s) diver (W) travels before stopping with (V_i) at the start and a final velocity of zero:

$$a_{decel} = \left(V_f^2 - V_i^2\right)/(2*s) = -V_i^2/(2*s)$$

The force on the diver due to this deceleration is:

$$F_{decel} = m*a_{decel} = (W/g)*V_i^2/(2*s)$$

What is causing this deceleration is the average drag force [1] on the diver needed to balance this force at distance (s) is:

$$F_{drag} = \frac{1}{2}*C_d*A*(\rho/g)*V^2 = 1.93*A*(V_i/2)^2 \text{ average drag force on flat surface in}$$
water since at equilibrium $F_{decel} = F_{drag}$ neglecting the effect of buoyancy that is small:

$$s = 2*W/(1.93*g*A)$$

Notice the velocity terms cancel as they are in F_{decel} and F_{drag}.

The displaced water (W_w) has to be moved from under the diver and will move sideways and up as is shown in Figure 5.8.2.

Since the pool is a fixed volume and neglecting the pool expanding radially elastically, W_w will probably splash out of the pool at V_i.

With $W = 200\,lb_f$, $A_{frontal} = 3\,ft^2$ (back arched), $h = 36\,ft$.

$$s = 2*W/(1.93*g*A) = 2.2\,ft$$

This is excessive as the pool depth was only 1 ft. The pool also expanded radially to absorb some of the energy which may account for some of the difference.

$$V_i = (2*g*h)^{1/2} = 48\,ft/s$$

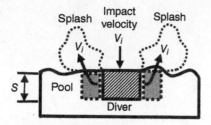

Figure 5.8.2   Water splashing out of pool.

The body experiences a short period of deceleration (G's);

$$G's = a_{decel}/g = -V_i^2/(2*s*g) = 17\, G's$$

This means each organ in the body will see 17 times its normal weight.

A very short-duration military aircraft seat ejection results in 25 G's and little physical injury but at 50'G is in the area of severe bodily injury [2]. Arresting straps such as used by window washers are designed for about 15 G's.

A normal arched back belly flop from 10 ft into a large pool would see about 5 G's. A graceful smooth entry dive from 10 ft with $A = 0.8\,ft^2$ would experience less than 2 G's.

A cliff diver at 92 ft would experience about 8 G's. Some say it feels like getting a body punch from a boxer.

5.8.1   Summary

Belly flop dives result in higher G's on the internal and external body parts. Along with displacing a foot of water, the flexible pool probably helped the diver survived by absorbing more shock, meaning it expanded radially. This wasn't considered in the analysis. In addition, the diver arched his back inward like a swan dive, as shown in Figure 5.8.1 so the initial frontal contact area was much less. This allowed the body to sink in further, increasing the absorbed energy. However, the real bravery comes in trying to find that tiny pool on the way down from 36 ft. Being off just a little and it wouldn't have been survivable.

References

1 Sofronas, A. (2006). *Analytical Troubleshooting of Process Machinery and Pressure Vessels*, 144. Wiley & Sons.
2 Crawford, H., *Survivable Impact Forces On Human Body Constrained By Full Body Harness*, HSL/2003/09. Crown Copyright 2003.

5.9 How Hard Is a Boxers Punch?

I recently saw boxing in slow motion on a video and it sure looks violent. Since I was investigating impacts to the head I wondered what G force a punch would create. The effect of G forces depends on the amount of time during which it occurs. A single 50-G football impact might take a fraction of a second and the brain can handle it. Extend that to a minute such as in an acceleration test and the person would die due to a lack of blood flow to the brain. So a sudden G impact causes brain injury by brain mechanical damage and a longer term G force is from blood depravation. A punch would fall into the former category.

Figure 5.9.1 is the actual model analyzed and is similar to that used in Section 5.4.

When the hand hits the head, the two masses join together but the velocity of the larger head and upper torso (V_2) will be less than the velocity of the smaller arm and hand (V_1). The energy is absorbed by the body spring. The spring constant (k) is the horizontal deflection (δ) of the head and a portion of the torso against the neck due to a force (F) and is what restrains the head.

Equate the momentums assuming the masses stick together and don't bounce:

$$m_1 V_1 = (m_1 + m_2)*V_2 \text{ then } V_2 = [m_1/(m_1 + m_2)]*V_1$$

The kinetic energy becomes:

$$KE_2 = (\tfrac{1}{2})*m*V^2 = (\tfrac{1}{2})*(m_1 + m_2)*([m_1/(m_1 + m_2)]*V_1)^2$$

The potential energy PE_s of the spring (k) is:

$$PE_s = (\tfrac{1}{2})*k*\delta^2$$

Equating KE_2 to PE_s as the spring absorbs KE_2 and solving for the displacement of (m_2):

$$\delta = [m_1/(m_1 + m_2)]*V_1*[(1/k)*(m_1 + m_2)]^{1/2}$$

The time (t) = 0.1 seconds is the time for the hand to move the distance L_{ft} = 2.5 ft so the initial velocity of the hand (m_1) from the punch is:

$$V_1 = L_{ft}/t = 25 \text{ ft/s}$$

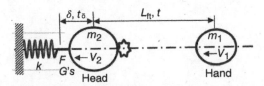

Figure 5.9.1 Lumped mass model.

The force on the head is:

$$F = k*\delta$$

For the glove, hand and arm and $m_1 = W_1/g$. The mass of head, neck, and top part of the body $m_2 = W_2/g$, when $W_1 = 50\,lb$, $W_2 = 100\,lb$, $k = 1,000\,lb/ft$, and δ is calculated from the previous equation.

$F = 570\,lb$ force on boxers head.

That's a short-term impact and it probably hurts quite a bit because the face has a lot of nerves in it.

5.9.1   Summary

A punch is a short duration impact. The impact force of the punch in this case is about 600 lb.

Professional fighters have punches with a velocity of 25 mph and a force of 775–1,300 lb$_f$.

This model was used in engineering impact problems where a mass impacted a larger mass and the force required to do this was needed for stress calculations.

6

Gas Explosion Events

Analysis Is Like A Time Machine. It Lets You See The Past And The Future

6.1   Energy in Steam Boiler Explosions

Steam boiler explosions happen all over the world as an Internet search will reveal. As plant engineers and operators, we should respect their destructive potential and know why they occur.

A steam explosion occurs when some mechanism causes the shell of a vessel containing high-pressure water and steam to rupture.

To help in explaining this, a greatly simplified fire-tube boiler will be analyzed.

Figure 6.1.1 shows a fire-tube boiler that has hot gas inside the tubes instead of water.

The boiler containment shell holds the pressurized water that covers the tubes and above that is pressurized steam. At STP conditions, water boils at 212 °F. Adding more heat will just boil or nucleate it more violently, but the temperature remains at about 212 °F. With a safety relief on the boiler shell set at 15 lb/in.^{2}, the temperature will rise to 250 °F before boiling with the excess steam vented. This is what happens with a home pressure cooker. The pressure keeps the water from nucleating, meaning having the bubbles of water vapor escape, thus the higher temperature.

Consider that a shell of volume (V_s) contains a volume (V_w) of water that is being heated with (Q) BTU/h of heat. Assume that the shell ruptures for some reason. The pressure in the shell suddenly drops to atmospheric pressure, and the water boils violently. This is because the pressurized superheated liquid water expands in volume by over 1,500 times when it is suddenly released to atmospheric conditions in a fraction of a second. This is when the real damage is done and is called a steam explosion or BLEVE (i.e. boiling liquid expanding vapor explosion). Slugs of violently boiling water develop, and the high density of water and its velocity

Unique Methods for Analyzing Failures and Catastrophic Events: A Practical Guide for Engineers, First Edition. Anthony Sofronas.
© 2022 John Wiley & Sons, Inc. Published 2022 by John Wiley & Sons, Inc.

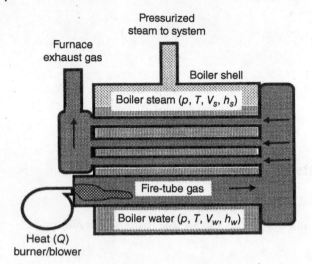

Figure 6.1.1 Fire-tube boiler vessel model.

result in immense forces. These forces can rip open the shell and propel fragments and even the boiler itself like a rocket. The exiting steam causes a pressure wave to develop and acts as a dynamic force on the surroundings.

Industrial boilers used in making steam can have these types of failures. There are many causes such as manufacturing defects, stuck relief valves, corrosion, or poor repairs. In any case, more heat (Q) into the boiler will cause a higher pressure if not vented.

Consider a small industrial boiler operating with the following conditions with $V_s = 60\,\text{ft}^3$ containing $W = 3{,}100\,\text{lb}$ water. A failure occurs at $150\,\text{lb/in.}^2$ where the water has a corresponding temperature of $366\,°\text{F}$ [1].

Enthalpy (h BTU/lb) is a property of a system. Here it is the internal energy of the water, such as heat plus the product of the pressure and volume. For water, its values are conveniently tabulated in steam tables. At the (p) and (T) conditions mentioned, this represents a liquid enthalpy of 338 BTU/lb. Saturated water at STP conditions is at 180 BTU/h. Consider that the energy available is the difference or 158 BTU/lb. Since there are 3,100 lb of water, the theoretical energy available is $4.9*10^5$ BTU or $3.8*10^8$ ft lb. Only a small portion of this energy, typically 10–25%, actually produces the BLEVE because not all of the 3,100 lb of water is heated evenly. It depends on the amount of heat energy put into it. The resulting energy, even 10%, is still massive, and that's why steam boiler explosions can be so devastating. A boiler destroyed a 500-ft^2 concrete block building with significant damage to adjacent buildings when the shell ruptured [2].

A similar failure occurred on a much larger boiler. This was also a 150-lb/in.^{2} boiler, but it contained about 15,000 lb of water heated to 366 °F. A low water level was suspected and resulted in a steam explosion with the theoretical energy being $1.8*10^9$ ft lb. This ruptured the end cap and sent the steam boiler rocketing through a concrete wall [3], moving it about 100 ft from its mounting. It weighed about 20,000 lb without the water.

On a more personal note, the power of steam in a home water heater explosion was noted in reference [4]. All of the water heater safeties were by-passed, and the pressure was allowed to build to 330 psig. The water heater went through a floor and a roof of an experimental house. Quite devastating.

6.1.1 Summary

No attempt will be given here on how to prevent such failures from occurring as there are just too many possibilities and too many different scenarios. Attending training seminars prepared by well-known societies, such as ASME, recognized boiler code inspection agencies, or learning about the equipment, controls, and operation for the specific unit used from the manufacturer would be most beneficial. Maintaining current code requirements, inspections, and testing is usually mandatory.

References

1 Crane Company. (1980). Flow of fluids. *Technical Paper No. 410.*
2 Internet Search. Rako, J., Gao, J., Into The Red Zone, Boiler and Plant Accidents, B.C.I.T. Boiler Explosion.
3 Internet Search, State Of Tennessee, Division Of Boiler And Elevator Inspection, Boiler Accident Dana Corporation, Paris Extrusion Plant, 2007.
4 Internet Search, MythBusters, Exploding Home Water Heater, Episode 89, 2007.

6.2 Delayed Fireball-Type Explosions

When a vessel containing a flammable liquid under pressure, such as in an LNG road tanker truck, ruptures and ignites, a vapor fireball explosion can occur. An internet search shows that many such explosions occur worldwide.

The purpose of this section is to provide safety information to those involved with such products.

The key point is that these types of explosions may not occur immediately. A surrounding fire can weaken the metal to the rupture point causing a delayed reaction. The fireball explosion can happen instantaneously or several minutes after being engulfed in flames. Lethal radiated heat can be present within 200 m (650 ft) or more. This makes rescues and extinguishing such fires complex for emergency responders. Professional training in what to do under these situations is always recommended [1]. Calculations will not provide the answer since the science is not exact and the amount of product in the tanker truck is usually unknown.

The following analysis helps in illustrating the effect of the many variables and dangers involved.

The random flying fragments are always a concern; however, the lethal thermal radiation effect of the fireballs is extensive.

How far away is a safe distance? Flying fragments from pressure effects may not have a reasonable safe distance. Some sources [2] mention that 80% of the debris lands within $4*R_{max}$, and in rare instances, up to $30*R_{max}$, where R_{max} is the calculated maximum radius of the fireballs. The danger from the heat radiation of the fireballs can be estimated for such blasts. Table 6.2.1 shows the damaging effect of this heat radiation energy on a surface area. Notice that this is for a 10-second exposure. Longer exposures would result in a heavier dose and shorter a lesser dose. Duration of 10 seconds is typical for the mass explosions of a tanker truck.

Consider that an LNG tanker truck with a load of $M = 19,000$ kg of propane (10,000 gal) overturns, a fire erupts, and the tanker explodes into a fireball after 10 minutes engulfed in a fire. The heat of combustion (H_c) for propane is 50,000 kJ/kg.

From experimental data, the fireball duration (t) is approximately,

$$t = 0.45*(M)^{1/3} = 12 \text{ s}.$$

At the end of the fireball growth period (t), it achieves its maximum radius,

$$R_{max} = 2.9*M^{1/3} = 2.9*(19,000 \text{ kg})^{1/3} = 77.4 \text{ m (252 ft)}$$

Table 6.2.1 Effect of surface heat flux [3].

Surface heat flux (kW/m²)	Effect of 10-s exposure
200–400*	Thin aluminum melts, wood, paint char. Total destruction
Above 70	Usually lethal
30–70	Second-degree burns to lethal
5–30	Pain and blistering, probably OK protective gear
1	Solar radiation

* Maximum at fireball surface.

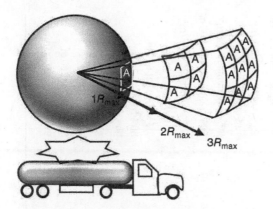

Figure 6.2.1 Fireball effect at a distance.

The surface of a sphere where the vapor has the correct air/fuel mixture and starts to burn has an area of,

$$A = 4*\pi*R_{max}{}^2 = 4*\pi*(77.4)^2 = 75{,}282 \text{ m}^2$$

Some experts [4] say that only 25% of the energy goes into heat radiation and the rest into pressure development and other sources. The radiation energy on the surface of the fireball is,

$$I = E/A = 0.25*50{,}000 \text{ kJ/kg}*19{,}000 \text{ kg}/75{,}282 \text{ m}^2 = 3{,}155 \text{ kJ/m}^2$$

In terms of power, the surface of the fireball would be,

$$P = I/t = 3{,}155 \text{ kJ/m}^2/12.0 \text{ s} = 263 \text{ kW/m}^2$$

This radiant heat effect at ground level, at some distance $x*R_{max}$, neglecting atmospheric conditions, can be estimated.

As the distance from this surface increases as shown in Figure 6.2.1, the surface intensity is decreased by the square of the distance. For example, when you are $3R_{max}$ away from the surface density of area, the (A) decreases by nine times because of the geometry.

For this example, at a distance $2R_{max}$ or 155 m (504 ft),

$$P = 263*\left[(1*R_{max})^2\right]/\left[(2*R_{max})^2\right] = 66 \text{ kW/m}^2$$

From Table 6.2.1, this amount of radiated heat would be lethal. Being hit by random flying fragments is a possibility even at a distance of $30*R_{max} \cong 2{,}300 \text{ m}$ (7,500 ft).

Table 6.2.2 compares some observations of actual tanker fireball explosions with the calculation methods shown. It can be compared with Table 6.2.1, which helps in validating some of the assumptions made.

Table 6.2.2 Historical data and calculated.

Accident	Amount involved (M)	Observed at crash site	Calculated
1972 Route 501, Lynchburg, VA, USA Road tanker explosion	8,800-kg LPG	Several minutes engulfed Fireball about 60-m radius At 80-m fatality At 125-m serious burns	R_{max} = 60 m, 263 kW/m^2 135 kW/m^2 at 80 m 60 kW/m^2 at 125 m
2002 Catalonia, Spain road tanker explosion	19,000-kg LNG	Fireball after 20 min At 200-m burns Melted aluminum frags. 260 m	R_{max} = 77 m, 263 kW/m^2 36 kW/m^2 at 200 m 80% fragments to 310 m 99% fragments to 2,300 m
2010 Port-la-Nouvelle, France Road tanker explosion	4,300-kg LNG when 65%	Engulfed 20 min Total destruction 20 m At 200-m first responders OK Frags. to 150-m broken windows to 700 m	R_{max} = 47 m, 263 kW/m^2 14 kW/m^2 at 200 m 80% fragments to 200 m 99% fragments to 1,400 m
2011 Zarzalico, Spain road tanker explosion	21,600-kg LNG After spill 12,000-kg remaining	Firefighters at 200 m Persons at 600 m felt warm air Frags. to 200 m	R_{max} = 66 m, 263 kW/m^2 30 kW/m^2 at 200 m 1.3 kW/m^2 at 600 m 80% frags. to 260 m 99% frags. to 2,000 m
2012 Kerala, India Road tanker explosion	19,000 kg LPG. Probably 12,000 kg left after leaking	Engulfed for 25 min Destruction to 15-m radius At 200-m burns Fatalities within 200 m Large metal pieces 400 m	R_{max} = 66 m, 263 kW/m^2 30 kW/m^2, 200 m 80% frags to 260 m 99% frags. to 2,000 m

6.2.1 Summary

The results of Table 6.2.2 indicate three important facts:

(1) The fireball-type explosion can occur at any time when containers are engulfed in a fire.
(2) The thermal radiation can be lethal within 200 m (650 ft) and beyond and depends on the mass (M) in the explosion.
(3) The range of debris from the explosion can be 400 m (1,300 ft) and beyond.

This shows the importance of evacuating such accident areas quickly and having the first responders well trained in what to do during these types of hazardous events [1].

References

1 (1999). *Center for Chemical Process Safety of the American Institute of Chemical Engineers, Guidelines for Consequence Analysis of Chemical Releases.* ISBN: 0-8169-0786-2.
2 Roberts, A.F. (1982). Thermal radiation hazards from releases of LPG from pressurized storage. *Fire Safety Journal* 4.
3 Hymes, I. 1983. The physiological and pathological effects of thermal radiation. United Kingdom Atomic Energy Authority, SRD R 275.
4 Zhang, Q. and Liang, D. (2013). Thermal radiation and impact assessment of the LNG BLEVE fireball. *Procedia Engineering* 52.

6.3 Method for Investigating Hydrocarbon Explosions

Vapor explosions in the hydrocarbon processing and the transportation industries occur all too often.

A detailed investigation is required to determine the cause or causes. Specialists may be requested to provide additional information to help the investigating team verify the source and magnitude. This can be done by analyzing the fire, blast pressure, and debris field damage data and working back to the explosion source and magnitude.

A simple mathematical technique explaining this is presented here for educational purposes.

This type of analysis is complex, and even with sophisticated computer programs, experience is required when interpreting the data.

Explosions are characterized by the amount of hydrocarbon product involved in the initial explosion and the resulting damage. This results in the following:

- Thermal radiation created by the fireball. The fire damage can be used to approximate the source and magnitude.
- Overpressurization from the blast wave. Structure damage and glass breakage can be used to define the source and magnitude.
- Flying fragments due to the blast wave. The location of the pieces can be used to estimate the source and magnitude.

While a steady-state flame may burn for hours, it is the initial explosion that causes the most damage. Figure 6.3.1 represents a plan view on distance from the source information that can be obtained by modeling the explosion event. It can also provide estimates on the magnitude of the explosion and size of the rupture. This analysis is based on a symmetrical explosion. Nonsymmetrical explosions as shown can occur based on how the explosion is directed. In these cases, distances can exceed those of a symmetrical explosion (Table 6.3.1).

The empirical method presented here has been compared to many industrial explosion investigation reports and results in reasonable answers.

Explosions are caused by volatile hydrocarbon vapors from leaks or ruptures. When ignited, a shock wave and fireball-type explosion occurs.

This analysis considers a hole or rupture in a pipe or vessel which is under pressure and flows out for some time. This defines the amount of product involved in the initial explosion (Eq. (6.1)). Likewise, the volume of contained gas, if known, can be used.

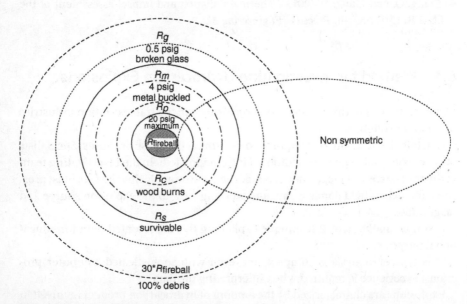

Figure 6.3.1 Explosion zone and effect.

Table 6.3.1 How the data is organized.

Designation	Calculation basis	What Is observed and collected
R_{fireball}	Fireball radius based on amount of energy release and other factors up to 35 BTU/s/ft^2	Fireball visual observation if available
R_p	Destruction of most buildings based on 10–20 psig and distance from fireball	Total destruction of wood, metal to concrete buildings
R_c	6 BTU/s/ft^2 energy to burn wood and distance from fireball	Trees and sides of wooden building or telephone poles scorched
R_m	Metal building buckled Based on 4-psig pressure and distance from fireball	Roofs, walls, and doors blown in
R_s	Survivable from burns based on distance of radiation from fireball 1 BTU/s/ft^2	Survival of exposed humans and animals with burns and no protective gear
R_g	Broken glass from 0.5 psig and distance from fireball	Plate glass windows of homes broken
R_d	Maximum expected debris field ($30*R_{\text{fireball}}$)	Small pieces of metal will be found within this radius

Bernoulli's equation can be used to estimate the flow (Q_{gas}) out of a diameter ($d_{\text{in.}}$).

$$Q_{\text{gas}} = 13.27*d_{\text{in.}}{}^{2}*(\Delta p_{\text{psig}})^{1/2} \text{ ft}^3/\text{s}$$

The pressure differential is Δp_{psig} to atmosphere.

One way to use the analysis is to assume a rupture diameter ($d_{\text{in.}}$) and compare the results with the visual damage data. When they agree, this diameter will represent the apparent size of the leak that caused the explosion. This is usually necessary as the explosion may have resulted in larger secondary ruptures. It also approximates the energy in the explosion (Eq. (6.2)).

The following analysis calculates the fireball radius (R_{fireball}). Within the fireball radius maximum damage occurs. Aluminum can be expected to melt, steel bends, and the pressure blast can be much higher than 20 psig. The critical radius (R_c) is defined as the zone where the radiation intensity causes wood to burn and paint to melt. Within a given radius (R_m), damage from the blast pressure wave will be strong enough to buckle sheet metal structures and knock people off their feet. Within R_g, home window glass will break. Over 80% of all flying fragments will be located within $4*R_{\text{fireball}}$.

Since the duration of the vapor cloud leak is seldom known, the time for it to develop into a combustible mixture and burn in a large explosion is estimated as $t_s = 10$ seconds based on actual data. For instantaneous rupture of vessels, all the remaining weight of the product in the vessel is used for energy calculations. When there is a long-term leak as in Case 1, the duration time is used in (Eq. (6.1)), not 10 seconds.

The weight (M) of combustible mixture is:

$$M = \rho_{gas}*Q_{gas}*t_s \text{ lb} \tag{6.1}$$

When a volume is known such as a room full of a combustible mixture,

$$M = \rho_{gas}*V_{cuft} \text{ lb}$$

The radius of the fireball,

$$R_{fireball} = 7.2*M^{1/3} \text{ ft}$$

Surface area of fireball is:

$$A_{fireball} = 4*\pi*(R_{fireball})^2$$

A conservative value for the heat of combustion (H) of gases such as propane, propylene, and natural gas is:

$$H = 21,500 \text{ BTU/lb}$$

The total energy involved in the explosion is Eq. (6.2),

$$E_o = H*M \text{ BTU} \tag{6.2}$$

Only 25% of this energy goes into generating the fireball, and the rest goes into the pressure wave and other factors. The heat flux (I) is:

$$I = 0.25*E_o/A_{fireball}$$

Radiation intensity (P) where $t_s = 10$ seconds is the typical time for a vapor cloud to achieve an explosive mixture and burn is,

$P_{fireball} = I/t_s$ on the surface of fireball is usually limited to a maximum of 35 BTU/s ft^2

This diminishes beyond the fireball radius,

$$R_{critical} = R_{fireball}/(P_{critical}/P_{fireball})^{1/2}$$

The value $P_{critical} = 6$ BTU/s ft^2 is the point that wood burns at 800 °F. Within this zone, severe heat damage can be expected.

Tests show that 80% of flying fragments are located within $4*R_{fireball}$, but small pieces can be up to $30*R_{fireball}$.

The pressure wave can be approximated at a distance away from the fireball center, using the following method [2].

The value $\eta_b = 0.5$ is the fraction of total energy used in the blast wave, and R_s is the distance of interest.

$$R = 0.48*R_s/((\eta_b*E_o)^{1/3})$$

For R less 0.3 use,

$$\Delta p_{psig} = 0.75/R^{1.7} \text{ psig}$$

For R greater than 0.3 use,

$$\Delta p_{psig} = 0.87/R + 0.20/R^2 + 0.36/R^3$$

Example 6.1 The following example illustrates the use of the method in a catastrophic explosion at a manufacturing plant in Houston, Texas, on January 2020.

From news coverage, drone footage, data communicated by observers, and overlaying this information on a Google Earth map of the site, an idea of the damage can be approximated. The preliminary investigation indications were that it was a pipe leak from a 2,000 gal propylene container over a 24-hour period. The mixture was ignited, possibly by an electrical spark. The tank was in a building with a volume of about 80,000 ft^3, so it was released in a confined area that could have slowly filled with the leaking gas to a localized explosive mixture. This is the volume which was considered filled with an explosive mixture.

Table 6.3.2 represents the observed and calculated results for this case.

Assuming the leak time the volume of gas was released from a small hole, calculations can be performed until they closely match the observed damage data. After matching the following results were available. The destruction was centered on the building in which the tank was stored. It would be similar to that of a 0.125-in.-diameter hole or rupture area in a 250-psig vessel leaking for 8 hours before ignition.

These types of analysis have limited value on their own because of all the assumptions and observations made. They can help organizing the data for the investigation team for a better understanding of the cause.

Table 6.3.2 Observed and calculated destruction.

Drone and reported observation of destruction radius	Calculated (ft)
No data	$R_{fireball} = 150$
80 ft – Building totally obliterated	$R_p = 200$
No data	$R_c = 250$
150 ft – Person reported survived	$R_s = 600$
500 ft – Badly buckled building metal walls	$R_m = 350$
1,000 ft – Structural damage homes and broken glass	$R_g = 2,000$
2,000 ft – Debris found	$30 * R_{fireball} = 4,600$

Case History 6.1 [3]:

The shaft (3.75-in. diameter) failed when one side blew out. In this case, the duration of the escaping gas was recorded as four minutes and only one explosion occurred. Operators 100-ft away say that they saw a drifting vapor cloud of about 15 ft high. The explosion tore off their protective gear and threw them several yards.

The report states the leak was only for 4 minutes, windows broke at 3,200 ft, people were knocked over by the blast at 300 ft, sheet metal buckled 500 ft from the blast, and there was extreme fire and pressure damage 150 ft from the blast center. The blast was asymmetric and occurred in a highly congested area, and it was the reason for lack of distant debris and personnel survival at 100 ft from the explosion. A vessel explosion, which this wasn't, usually has a large number of fragments associated with it.

Table 6.3.3 represents the calculated and observed results for Case History 6.3.

Table 6.3.3 Case history calculated and observed results.

Drone and reported observation of destruction radius	Calculated (ft)
150 ft observed	$R_{fireball} = 190$
150 ft – Building totally obliterated	$R_p = 170$
150 ft – Extreme fire damage	$R_c = 360$
100 ft – Person reported survived	$R_s = 800$
500 ft – Badly buckled building metal walls	$R_m = 350$
3,200 ft – Structural damage homes and broken glass	$R_g = 2,000$
2,000 ft – Debris found	$30*R_{fireball} = 5,700$

Case History 6.2 [4]:

This tragic explosion occurred in Beirut, Lebanon, during the writing of this book. It involved the explosion of a stated 2,750 tons of stored ammonium nitrate, used in fertilizers and explosives. Some data was available on the extent of the property damage. The human suffering was immense.

Visible fireball radius size from distant videos appeared to be in the 800-ft range when compared with the grain silos length of 500 ft It seems to have occurred much faster than a gas cloud explosion. The disastrous effect seems to be similar.

For this analysis, the energy $H = 2,200$ BTU/lb was used as a TNT equivalent with 2,000 tons involved.

Table 6.3.4 represents the calculated and limited observed results for Case History 6.2.

Table 6.3.4 Beirut explosion observed and calculated.

Reported observation of destruction radius (4, 5)	Calculated
Observed radius fireball 800 ft	$R_{fireball} = 1,100$ ft
Probably at least 1,000 ft	$R_c = 1,600$ ft
2,000 ft – Buildings total destruction	$R_p = 1,500$ ft
3,000 ft – St. George Hospital badly damaged	$R_m = 2,000$ ft
2,000 ft – Survivable but sheltered	$R_s = 4,000$ ft
5 mi – Reported airport damage ceilings, windows	$R_g = 2$ mi
Debris not reported but all debris could be in $30*R_{fireball}$	$30*R_{fireball} = 6$ mi

The extent of the limited data for Case History 6.2 suggests that the TNT equivalent of 2,000 tons would cause such destruction. This is about equivalent to the stored 2,780 tons of ammonium nitrate and is greater than any nonnuclear weapon available. The crater diameter was 400 ft about the size of the Oppau [6] explosion that contained ammonium nitrate which had a TNT equivalent of 1,000–2,000 tons.

Case History 6.3 [7]:

Several years ago overpressurization of the supply gas to homes caused explosions. This resulted in a large fireball and total destruction of homes and major damage to neighboring homes. To explain the magnitude of energy into an explosion that we can visualize, meaning the total destruction of a home, the following analysis is given.

In this analysis, it is assumed that a typical kitchen where the explosion occurs has a volume of 2,000 ft^3. The density of the gas times this volume will be assumed to be the explosive mixture involve. The procedures used to calculate fireball size and distances is the same as have been used previously.

(Continued)

Case History 6.3 (Continued)

The approximate size of the fireball could be estimated by videos of the explosion and comparing them with the size of the house. It was evident from the videos that neighboring homes were also badly damaged (Table 6.3.5).

Table 6.3.5 Home explosion observed and calculated.

Reported observation of destruction radius	Calculated (ft)
60 ft – Fireball radius from videos	$R_{fireball} = 45$
80 ft – Home totally obliterated	$R_p = 50$
70 ft – Side nearby home scorched	$R_c = 50$
No data	$R_s = 100$
No data	$R_m = 125$
No data	$R_g = 500$
No data	$30*R_{fireball} = 1{,}300$

This explosion was equivalent to about 0.5–1 ton of TNT depending on the energy into the explosion. Home owners were instructed to shut off the gas supply to their homes until the cause could be determined and rectified.

6.3.1 Summary

Mathematical analysis techniques can be used to help defining the cause of a gas cloud or TNT equivalent explosion. The more on-site data collected, the better the results. The examples shown here only contained data obtained from reports. Temperature and blast waves from the explosion result in damage and debris that can be used to help defining the source and cause. There appears to be no realistic safe distance from such explosions, and first responders need to be well trained and prepared.

References

1 Sofronas, A. (2019). Case 106: Delayed fireball type explosions with references. *Hydrocarbon Processing Magazine*.
2 Dong, S. et al.. Full scale method scale experimental verification of the explosion shock wave model of a natural gas pipeline, https://doi.org/10.1155/2018/4202389.

3 EPA/OSHA Joint Chemical Accident Investigation, EPA 550-R-96-005, June 1998.

4 New York Times, Beirut Explosion, August 4, 2020.

5 Honeywell, E., NASA Maps Beruit Explosion Damage From Space, Space.com

6 Wikipedia On Line Information, Oppau Explosion, 2019.

7 Wikipedia On Line Information, Merrimac Valley Gas Explosions, 2018.

6.4 Pipeline Explosion Critical Zone

This shows a pipeline explosion in Kentucky. It was analyzed in the same manner as in Section 6.3. The importance is to show the nonsymmetric-type affected blast zone. The pipe rupture due to damage caused it to be forced out of the ground and pointed in the trench direction shown in Figure 6.4.1 as was most of the destructive energy.

This illustrates one of the difficulties with trying to predict the effect of an explosion of any type. The major damage prediction is never really possible since the direction the blast will take is never predictable. The rupture, terrain, or other barriers may direct the blast energy in various directions.

6.4.1 Summary

Explosions can have unexpected effects. Distance from an explosion is the only safe measure.

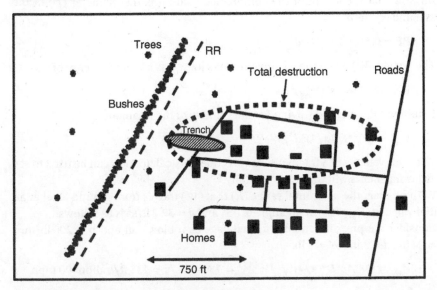

Figure 6.4.1 Pipeline nonsymmetric destruction zone.

6.5 Pneumatic Explosion Debris Range

A vessel was to undergo a pneumatic pressure test. Such tests are used to establish the integrity of a repaired vessel. The group performing the test said an exclusion zone at a 100-ft radius from the test site would be acceptable. The following calculations were used to show this wasn't adequate.

Usually the test is a hydro-test because of the much lower energy involved. This energy effect can be observed with a simple test. If we stick a pin in a water-filled balloon and one filled with air, one will just leak and the other will pop. In real life, a large vessel under high pressure won't just pop but explode. This occurred in Shanghai in March 2018 when a system was pneumatically pressure tested to 2,200 lb/in.^{2} and exploded, and half of the 1-acre unit was destroyed. The fatal debris piece was found 1,200 ft from the explosion site.

A pressurized column of gas compressed like a spring and acting on an area (A) is potential energy (PE). When the area suddenly breaks out, it converts into kinetic energy, which represents the danger. This is what happens with a potato cannon.

The amount of PE in a volume of pressurized air is like compressing an air spring,

$$PE = 0.5*k*\delta^2$$

$k = p*(\pi/4)*d^2/\delta$ where $(\pi/4)*d^2$ is the area that breaks out and column diameter (d).

Assume that the length of the compressed column is the diameter (D) of the vessel and equals δ,

$$PE = 0.5*p*D*(\pi/4)*d^2$$

The weight (W) when the fragment breaks loose has a kinetic energy of,

$$KE = 1/2*W/g*(V_{avg})^2$$

Equating PE = KE and solving for the velocity of the fragment:

$$V_{avg} = (g*p*D*(\pi/4)*d^2/W)^{1/2}$$

With this velocity of a fragment, simple trajectory calculation can be used to see how far fragments will travel horizontally.

With no drag, the horizontal range (R) that (W) travels from ground level at an initial velocity (V_{avg} ft/s) at an angle of 45° and $g = 32.2$ ft/s^2 is as follows.

Consider the pressure on an attached gage as it blows off and $p = 200$ lb/in.^{2}, $D = 24$ in., $d = 3$ in., $W = 5$ lb.

$$V_{avg} = (386*p*D*(\pi/4)*d^2/W)^{1/2} = 1,620 \text{ in./s} = 135 \text{ ft/s about 90 mph}$$

$$R = (V_{avg})^2/(2*32.2) = 280 \text{ ft}$$

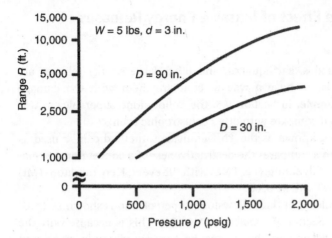

Figure 6.5.1 Range of fragments.

When the velocities are over sonic or about 1,200 ft/s, the velocity is probably limited to this value because of choked flow out of the opening.

Figure 6.5.1 shows the range under several conditions and (W) and (d) known. The size shown might be a gage or plug threaded into a pressure vessel.

Here is the major problem with trying to define a safe distance. Fragments of different sizes and weights from different volumes and pressures will have different ranges. Stating a safe distance isn't possible unless you know these variables.

6.5.1 Summary

This type of data is useful when recommending alternatives to pneumatic testing such as:

- segmenting piping into manageable low energy volumes.
- using hydro-testing with the proper precautions.
- testing only the weld areas repaired using temporary localized weld-on caps.
- using nondestructive testing methods of inspection (NDT).
- using a combination of the above.

We sometimes hear "The vessel is operated with gas at this pressure anyway, so what's the concern about testing?". One answer is that you don't want a massive failure in the testing phase either.

National codes and standards may be referring to blast wave damage safe distances, and not flying fragments. Speak up when someone says, "I think 100 ft is a safe barrier distance". This is not usually the case. Conventional wisdom says, stay as far away as possible and be behind a concrete barrier. Realize that shock waves can go over and behind barriers too.

6.6 How Are the Effect of Massive Energy Releases Compared?

When explosions, tornados, earthquakes, and other large energy releases are involved, it is beneficial to have a way to compare them with the damage they might do. For example, in Section 9.5, the Yellowstone supervolcano was investigated. How does it compare with other catastrophic events?

A comparison method known as the TNT equivalent method can be used as shown in Table 6.6.1. This compares the destructiveness of a conventional explosive device such as tons of dynamite (i.e. TNT) with the event. Here megatons (Mt) or millions of tons of TNT will be used.

Notice that the last release of the Yellowstone supervolcano is shown as $1*10^6$ Mt, but the analysis in Section 9.5 is about $1*10^4$ Mt. This is because only the rock ejecta was considered and not the magma, heat energy, vibration, and sound energy.

Table 6.6.1 Megaton equivalents.

Megaton	Event
0.002	Beirut, Lebanon 2020 explosion destroyed $1.0\,mi^2$ Magnitude 4.0 earthquake on Richter scale
0.01	Small atomic bomb completely destroys $100\,mi^2$
0.02	Magnitude 6.0 earthquake on Richter scale
1	Small nuclear bomb
3	All the explosives used in World War Two
10	Meteorite impact Meteor Crater Arizona 0.75-mi-diameter crater 50,000 years ago
12	E5 tornado Moore, OK 2013, 215 mph, destroyed 2,500 homes
15	Meteorite air explosion in Siberia in 1908 flattened 1,000 square miles and 80 million trees
24	Eruption total debris of Mt. St. Helen in 1980 in Washington
50	Magnitude 8.5 earthquake on Richter scale
50	Largest nuclear bomb (i.e. Tsar) ever detonated in 1961 in Siberia. Total destruction is of $1,000\,mi^2$
200	Eruption of Mt. Krakatoa in Indonesia in 1883 island, 70% eliminated
$5*10^3$	Energy release typical hurricane duration
$1*10^6$	Last eruption of Yellowstone supervolcano 640,000 years ago
$1*10^8$	Chicxulub impact by asteroid 66 million years ago. Mass extinction event and thought to have led to extinction of the dinosaur

6.6.1 Summary

The equivalent of millions of tons, megatons, of TNT explosives is used as a comparison on the severity of explosive and extremely high-energy events. One megaton = $3*10^{15}$ ft lb$_f$.

6.7 Engine Air Intake Manifold Explosion

This is a failure that occurred on an 800-HP integral gas engine-compressor that was located in a compressor house with 12 other engines. There were six of these types of engines that were pumping a refrigeration gas. One of the engines had an explosion in the intake manifold, which was only supposed to be handling pressurized air from a turbocharger. This is shown in Figure 6.7.1.

Figure 6.7.2 is a similar engine with an intercooler. The turbocharger is before the cooler. Notice that sketches don't look like the actual equipment but are simplified for a better understanding of the analysis. Figure 6.7.3 is a view of the compressor house which the engine was in.

The elbow where the explosion occurred was cast iron and shattered shrapnel within 20 ft from the explosion zone. No one was injured, but the potential was great. The six engines were shut down until the cause could be determined. This was a major production loss as the reactors required these engines for the polymer process.

The cause of the gas in the air intake manifold was due to the erroneous hook-up of a purge line. The ignition source was a carbon hot spot on this two-cycle engine.

An important question was why didn't the pressure relief valve limit the pressure rise. The valve was designed to relieve at 10 psig. Here's the analysis that was done and the explanation.

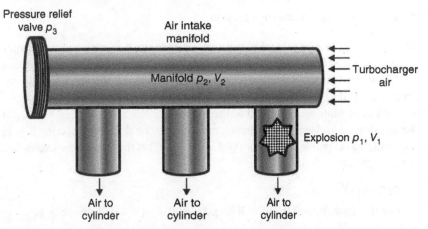

Figure 6.7.1 Air intake manifold explosion.

Figure 6.7.2 Similar engine to that analyzed.

Figure 6.7.3 Compressor house engine was in.

The purpose of the pressure relief valve (p_3) was to limit excessive turbocharger pressure and was set at 10 psig. Gas explosions confined to a 1.2-ft³ volume can reach 113 psia and occur in about 0.08 seconds [1]. The manifold elbow was a complex rectangular design but is shown as cylindrical with $V_1 = 2.0$ ft³.

Since the explosion was confined to the elbow section and occurred rapidly, it can be considered an isentropic process, meaning no heat transfer occurs. It is at (V_1, p_1) and then expands in the manifold into (V_2, p_2) at the time of the explosion. The perfect gas law for this is,

$$p_2 = p_1/(V_2/V_1)^k$$

For this case, $V_1 = 2$ ft³, $V_2 = 8$ ft³, $p_1 = 100 + 14.6 = 115$ psia, $k = 1.3$

$$p_2 = 19 \text{ psia or 4 psig}$$

6.7.1 Summary

The pressure valve didn't open because the pressure was only $p_2 = p_3 = 4$ psig, and the valve was set to open at 10 psig. This was because the gas expanded and was attenuated in the manifold volume. The shock pressure occurred in the elbow. This is like blowing a puff of air into a bottle. The pressure in the bottle is less than your puff because of the volume. Also the explosion occurred so quickly that when it blew a hole in the elbow, it also relieved the pressure in the manifold.

Reference

1 Mikaczo, V., et al. Simulation of propane explosion in closed vessel, http:// annals.fih.upt.ro/pdf-full/2017/ANNALS-2017-3-05.pdf.

7

Vibration and Impact: The Cause of Failures

Judgment Errors Occur; They Shouldn't Occur Early In Your Career Or Consecutively

7.1   Investigating a Possible Cause for a Coupling Failure in a Centrifugal Compressor

Torsional vibration is the oscillatory twisting of the shafts and masses in a rotor or geared assembly. It's superimposed on the mean torque in the system. It vibrates with a frequency that can be externally forced at the natural frequency of the system. When the forcing frequency coincides with a natural frequency, large twisting amplitudes can occur within the system. That is what we examined in this section.

Torsional vibration failures rose by 35% during the past decade according to one source [1], possibly due to the introduction of variable frequency drives. While critical systems such as motor–gear–compressor units are usually well instrumented, a direct measurement of a torsional vibration problem can be elusive.

Torsional vibrations, because of their nature, are usually only noticed from external vibratory reactions at a gearbox, the sound of chattering gear teeth, or a shaft or coupling fatigue failure. Accelerometers and proximity pick-ups near gears and bearings can observe abnormal motion and accelerations but are not usually directly related to the torques and angular displacements resulting from torsional vibration. Torque meters or angular displacement torsiographs perform that function but are not normally used for continuous monitoring. A torsional vibration analysis is required to relate a one point measurement throughout the system.

The type of analysis shown here was used for preliminary screening purposes, not for design purposes. When detailed analysis and testing was required, engineering firms that perform this routinely are usually utilized.

Unique Methods for Analyzing Failures and Catastrophic Events: A Practical Guide for Engineers, First Edition. Anthony Sofronas.

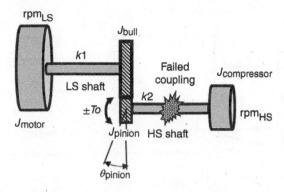

Figure 7.1.1 Motor–gear–compressor system.

Figure 7.1.1 illustrates a recent high-speed (HS) compressor coupling failure. This was an unusual coupling failure that occurred after 20 years of successful operation. The gears were found damaged, but they were left in and the HS coupling was replaced. The spare gears were being reconditioned. The new coupling failed again in a few days in the same manner with the damaged gears. A gear lubrication failure had caused the initial gear distress. Could this reused distressed gear have failed the high speed coupling in such a short time?

Being only an observer on this failure and not part of the investigation team, this analysis was an opportunity to gain valuable personal data for future use. For this reason, the actual mass elastic data was not known. Data from a similar system was used for this sensitivity study.

There are many failures of geared systems that can be analyzed using a force-damped torsional vibration analysis. Some examples are motor and compressor shaft and coupling failures, gear and teeth failures, and bearing failures.

Geared systems of this type are usually reduced to an equivalent three mass system for analysis purposes. Consider the system shown in Figure 7.1.2 where the masses and springs have been referenced to the motor side speed. The natural frequencies and relative mode shapes can be conveniently calculated in this way using various techniques.

The geared side looks different to the motor when this is done. For a speed increaser, the HS masses and spring constants appear much larger, which means they are harder to turn on the motor side for a speed increaser because of the mechanical advantage. The springs and masses on the high speed geared side are, therefore, multiplied by the square of the gear ratio. The system can be analyzed using a Holzer-type calculation on the first two modes of this three mass system, as discussed in Section 7.6. The first two relative mode shapes are shown in Figure 7.1.2. They are divided through by the largest (θ) value to normalize them to a value of 1 for clarity. This shows the nodes which have the highest torque and

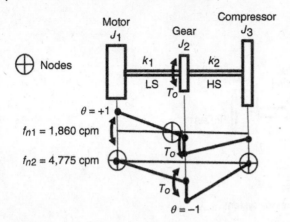

Figure 7.1.2 Equivalent system and mode shapes.

zero angular displacement (θ). For this analysis, notice that the highest torque is in the low-speed (LS) shaft not the HS for the first mode but in both for the second mode. The most sensitive point for a vibrating torque to act is at a point of maximum amplitude (θ).

The two calculated natural frequencies are f_{n1} (1,860 cpm) and f_{n2} (4,775 cpm). Campbell diagrams [2] are useful for determining where resonances will occur and if they will be a factor.

From Figure 7.1.3, it appears that the f_{n1} (1,860 cpm) is close to the operating speed and the 1× running speed would excite it. This is the most important for this study since the amplitude is greatest at the pinion where the exciting torque is acting. The f_{n2} (4,775 cpm) will be excited by the 3× running speed harmonic. Frequencies, sources, and magnitudes for these excitations are shown in Table 7.1.1.

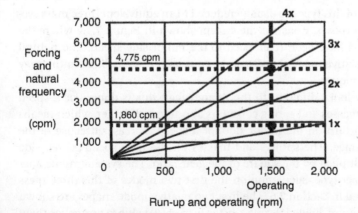

Figure 7.1.3 Campbell diagram.

Table 7.1.1 Vibration excitation sources and magnitudes.

Vibrating frequency	Source	Amplitude (T_o)
1×	Unbalance and misalignment also gear damage	±1% mean torque
2×	Misalignment also gear damage	±1% mean torque
All harmonics	Looseness, rattling, gear impacting	±1% mean torque
Single impact	Electrical spike	Several times mean torque

From Table 7.1.1, it appears that the first and second modes could be excited by the 1× and 3× operating speed. Damaged gears with impacting could excite both the 1× and 3× operating speed.

Failure stress diagrams for torsional vibration of shafts [3] suggest that cyclic vibratory stresses above ±5,200 lb/in.^{2} will usually result in fatigue failures. This means a crack probably started and grew until the coupling cross section couldn't handle the torque and twisted apart.

The concern is that both modes are operating near 1,500 rpm, especially if the mass elastic data is not accurate enough. For this analysis, it will be assumed that they are in resonance. This would only be relevant if something was exciting them, such as damaged gears. Since the unit ran for 20 years, the initial design with good gears had a very small exciting torque (T_o). The first failure also probably occurred shortly after the lubrication failure.

With the frequencies known, the vibration amplitude needs to be determined, so that the coupling tube stress can be calculated. A sensitivity analysis using 1% excitation (T_o) due to frosted distressed teeth will be used to determine the HS coupling stress.

Presented here is an approximate method for determining the actual torques and stresses in a geared system. This method has been used for analyzing many systems and seems to present results that agreed with the failures. The method allows historical experiences to be included, such as percent vibrational torque and damping ratios. It has been used on extruders, compressors, ships, aircraft, and fans driven by electric motors, steam turbines, and the most difficult of all internal combustion engines. This is because of the many harmonics and phasing in the firing sequence in these engines.

Knowing the system (ω_n) from the three mass equivalent model, interest is on the pinion gear being excited by gear damage. Consider Figure 7.1.4 with excitation torque (T_o) acting on the pinion because of damaged gears. Physically the pinion is forced back and forth through the clearance between the gear teeth, twisting the HS compressor shaft against the compressor [4].

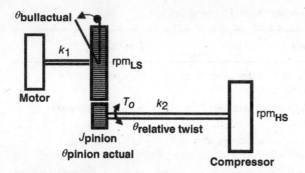

Figure 7.1.4 Vibratory torque acting on HS shaft.

The magnifier method [2] will be used to approximate the amplitude.

The magnification factor (M) is defined as the dynamic angular displacement (θ_{dyn}) divided by the static angular displacement (θ_{static}) of the system.

$$M = \theta_{dyn}/\theta_{static}$$

To determine (θ_{static}) angular displacement, hold J_{motor} and twist it. An approximation of the twist angle, based on a two mass system will be,

$$\theta_{static} = T_o/k_{Holzer}$$

The spring constant (k_{Holzer}) is established using the value of torque that statically twists the motor against the geared system and represents the Holzer spring constant.

For a system in resonance, $M = 1/(2{*}\zeta)$ where (ζ) is an experimentally determined damping constant.

Solving for the actual motion of J_{motor},

$$\theta_{motoractual} = M{*}(\theta_{static}){*}[J_{motor}/(J_{motor} + J_{driven})]$$

Excitation torque is also an experience-based value and is taken as some fraction (φ) of the mean torque at the critical speed being considered.

$$T_o = \varphi{*}T_{mean}$$

This $\theta_{motoractual}$ can replace $\theta = 1$ in the Holzer table and the natural frequency will not change.

The vibratory torque in the LS shaft is

$$T_{vibratoryLS} = \text{Holzer torque for motor-gear}$$
$$\text{shaft after replacing } \theta = 1 \text{ with } \theta_{motoractual}$$
$$T_{vibratoryHS} = \text{Holzer torque for gear-compressor}$$
$$\text{shaft after replacing } \theta = \theta_{motoractual}$$

This $T_{\text{vibratoryHS}}$ Holzer torque needs to be multiplied LS/HS to arrive at the actual value and the amplitude multiplied by HS/LS.

The general rule to get back to the actual system from the equivalent system is that for speed reducers: $\theta_{\text{actual}} = \theta_{\text{HS}}\text{*LS/HS}$, $T_{\text{actual}} = T_{\text{HS}}\text{*HS/LS}$ and for speed increasers:

$$\theta_{\text{actual}} = \theta_{\text{LS}}\text{*HS/LS}, T_{\text{actual}} = T_{\text{LS}}\text{*LS/HS}$$

The stress in each hollow shaft is,

$$\sigma_{\text{shear}} = 16\text{*}T_{\text{shaft}}\text{*}D_o / \left[\pi\text{*} \left(D_o^4 - D_i^4 \right) \right]$$

$$T_{\text{meanLS}} = \text{HP*63,000/rpm}_{\text{LS}} \text{ for motor side}$$

$$T_{\text{meanHS}} = \text{HP*63,000/rpm}_{\text{HS}} \text{ for compressor side}$$

The model is shown for a forcing (T_o) on the motor side bull gear. The pinion excitation is amplified on the compressor side.

Since this is a sensitivity analysis, ±1% on the bull will be used. In the case that this causes a failure at the HS coupling, it would show that poor gears could be the contributor. The damping factor is $(\zeta) = 0.1$ for the overall geared system [5].

Table 7.1.2 is a sensitivity study using the second mode model.

These results show that a ±1% excitation on the bull will cause a high vibratory shear stress in the HS coupling tube. When the vibratory torque is greater than the mean torque, the gears cycle and impact through the backlash clearance. This can cause gear chatter and impact loading. With good gears and only 1/10% excitation, there is no separation of teeth and a long life is expected, as had occurred.

Table 7.1.2 Sensitivity analysis damaged gears second mode.

Component	Vibration frequency (cpm)	Vibratory torque (in-lb)	Mean torque (in-lb)	$T_{\text{vib}}/T_{\text{m}}$ shaft	Vibratory stress (SCF = 1) (lb/in.²)	Mean stress (SCF = 1) (lb/in.²)
Compressor HS coupling, ±1% T_o on pinion, damaged gears, $\zeta = 0.1$, 4 in. dia. shaft tube 1/8 in. wall	4,775 second mode	±75,000	+20,000	3.7	±26,000	+7,000
Compressor HS coupling, ±1/10% T_o on pinion, good gears	4,775 second mode	±1,000	+20,000	0.1	±1,000	+7,000

This is due to the second mode. The LS shaft is more robust and has lower stresses.

Coupling stiffness values can vary by up to 30%, damping ratios by 50% especially on geared systems and the excitation torque, which means percent of mean torque is always questionable, so a high degree of accuracy shouldn't be expected. Even sophisticated forced-damped computer programs have these types of errors, and this is why experience and testing are so important. No matter which values are used, it's apparent from the analysis that gear damage can cause the HS coupling to fail in short order.

7.1.1   Summary

Frosted gear surfaces can result in high vibratory torques in sensitive systems as the $\pm 1\%$ of mean torque indicates. A $\pm 1/10\%$ mean torque value or better is what might be expected from a good tooth profile, well-aligned new gear set. In the case analyzed, the distressed gear teeth were enough to fail the new high speed coupling tube in a vibratory shear type failure with only three days of operation.

The system was a well-designed system from a torsional viewpoint, and only the gear lubrication problem that resulted in distressed gears was the cause.

This analysis doesn't provide the true cause of the failure. The real cause was the lubrication failure and reuse of the gears. This was not known until after the second failure. The analysis does provide information which would make the investigators feel that there is low risk in starting up with new gears and correcting the lubrication problem. After the first failure, there were many possible causes to consider and the analysis provided more data on one possible cause. Section 7.3 provides another example.

References

1 Connor, D.O., Gutierrez, J. *Reveal the Unmonitored Risks in Your Rotating Equipment-Part 1, Hydrocarbon Processing*, 2020.

2 Sofronas, A. (2012). *Case Histories in Vibration Analysis and Metal Fatigue*, 31. Wiley.

3 Wachel, J.C., Szenasi, F.R. (1993). Analysis of torsional vibrations in rotating machinery. *Proceedings of the 22nd Turbomachinery Symposium*.

4 Jackson, C., Leader, M.E. Design, testing and commissioning of a synchronous motor- gear-axial compressor. *Proceedings of the 12th Turbomachinery Symposium*.

5 Corbo, M., Malanoski, S. (1996). Practical design against torsional vibration. *Proceedings of the 25th Turbomachinery Symposium*.

7.2 Sudden Power Interruption to a System

Many times it's important to know the effect of instantaneous torque that is applied to a mass. This can determine if it was severe enough to initiate a failure. When the change in speed in a certain time period has been recorded the following approach can be used. One type of problem might occur during a power interruption and restart. Gearing, shafts, or couplings may have been damaged.

Before a model can be developed, it's important to understand what is occurring. Consider an automobile traveling at a certain speed, closely followed by a large truck which is traveling at the same speed. If the automobile driver suddenly jams on the brakes, the automobile will decelerate and the truck will impact it. The automobile doesn't hit the truck, the truck hits the automobile. This is called the braking inertia effect. How hard it hits depends on the time it takes. While this was all described linearly, it also applies to angular torsional problems.

Consider a motor rotating a pinion and a bull gear as shown in Figure 7.2.1.

When there is a change in speed of the motor large enough to separate the gear mesh through the backlash, for an instant, the speed of the pinion will be slower than the bull gear. This is because at that instant, they are independent systems, and the bull gear keeps rotating at the same speed, just like the truck analogy. The pinion force is not acting on the gear. This causes the bull gear, the truck, to run into the slower pinion, the automobile. Using Newton's second law this can be expressed in angular equation form, $T = J*\alpha$.

Consider that $\Delta\omega$ is the change in the velocity in rad/s and the time for the mass to decelerate is Δt.

$$\Delta T = J*\Delta\omega/\Delta t$$

Knowing these quantities, speed change $\Delta\omega = \Delta\text{rpm}/9.55$ in the instant Δt of time, the change in torque ΔT that this will cause can be approximated.

$$\Delta T = J_{bull}*(\Delta\text{rpm}/9.55/\Delta t) \text{ in-lb}$$

where J_{bull} in./lb/s^2, Δrpm rev/min, Δt = seconds.

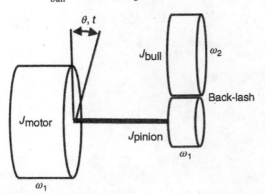

Figure 7.2.1 Suddenly applied torque to a mass.

Consider the following actual case. A power interruption occurred and a motor experienced a recorded speed change of Δrpm = 60 in. Δt = 0.05 seconds. The bull gear system had a mass of J_{bull} = 800 in./lb/s^2.

$$\Delta T = 800*60/9.55/0.05 = 100,500 \text{ in-lb}$$

In this case, $T_{mean} = 15,000$ in-lb, so $(T_{mean} + \Delta T)/T_{mean} = 7.7$. This is enough to recommend further investigation since a crack may have developed from this one-time event. In this case, the bolts clamping the disk pack coupling between the motor and pinion are sheared. Most shafts and gears are designed for a peak $(T_{mean} + \Delta T)/T_{mean} = 2$ which would include impact factors.

An inspection of the coupling and also opening the gearbox to inspect the teeth for any cracks would be justified as soon as possible. A cracked tooth eventually breaking off and falling into the gear mesh would result in a catastrophic failure.

Figures 7.2.2 and 7.2.3 represent the results of two electrical interruptions. Figure 7.2.2 is a coupling in a geared air compressor system which tried to restart as it was shutting down. The problem was with the shutdown logic. The separation of the disk pack from the overload is evident on both sides. The coupling bolts also sheared during the event. Figure 7.2.3 is a defect noticed on a gear tooth

Figure 7.2.2 Coupling failure due to overload.

Figure 7.2.3   Damaged gear tooth noticed after power grid interruption.

after a power grid interruption on a large-geared centrifugal compressor system. It is unknown if the outage was the cause.

This type of data, which means Δrpm and Δt_s, is usually only available on critical equipment; however, Δrpm = 3% in t_s = 0.05 seconds based on Figure 7.2.1 data, and other system averages can be used as an estimate when such data is unavailable.

7.2.1   Summary

Sudden power outages can result in high torsional impact loads in geared systems.

7.3   Effect of Liquid Slug in a Centrifugal Compressor

When machinery is out of service, it can be quite costly for a company, and there is significant anxiety to get the equipment back on-line. This becomes a major concern when the cause cannot be determined.

Failures where a cause hasn't been identified can't be rectified. This adds uncertainty to the start-up and with the continued operation of the repaired equipment.

Analysis can help when speculating on what might be the cause. Consider the failure of the HS disk pack coupling shown on Figure 7.3.1. Start-up dynamics, electrical surges, alignment, and compressor surging were not the problem. The failure mode of the coupling suggested a severe torsional overload failure. The centrifugal compressor was not damaged.

One thought was that liquid carryover (W_{ball}), sometimes called a liquid slug, could have been ingested by the centrifugal compressor. It is well known that reciprocating compressors fail with too much liquid ingestion since they are positive displacement machines and liquids are incompressible. Centrifugal compressors are not positive displacements, and the mechanism is different. The concern was the high speed of the impellers. A liquid slug might not show as damage to the compressor. Most of the literature just states no liquid allowed, or

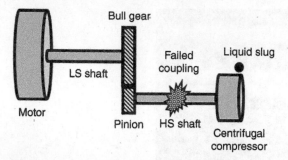

Figure 7.3.1 Centrifugal compressor system.

a catastrophic failure will occur with liquid ingestion. Centrifugal compressors are designed for 10-μm droplets of liquid and larger droplets result in erosion of the blades. At a particular size, major damage can be expected. Polymer balls of about 1-in. diameter and about the same density as water were known to have repeatedly bent the tip of an aluminum impeller with a tip speed of 5,600 in./s [1].

The following analysis will determine if a liquid slug was a viable concern in the coupling failure.

Consider the impeller shown in Figure 7.3.2.

In the model of Figure 7.3.2, an idealized sphere of liquid (W_{ball}), which means a slug, of a particular diameter (d) impacts a blade of an impeller (D). The reaction of the impact force (F_{impact}) is shown perpendicular to the blade which is rotating at a particular speed (rpm). The velocity of impact on the liquid ball is the impeller tip velocity (V_{tip}).

This sudden impact force on the impeller blade results in a torque (T_{impact}) on the coupling.

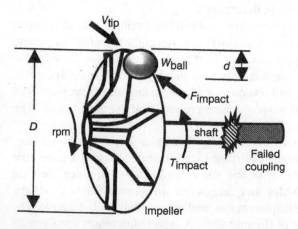

Figure 7.3.2 Impeller impact model.

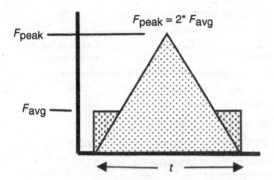

Figure 7.3.3 Peak impact force versus average.

Using Newton's second law or by equating momentum to the impulse force,

$$I = F*t \text{ impulse}$$

$$L = m*V_{\text{momentum}}$$

$$F_{\text{impact}} = (W_{\text{ball}}/g)*(V_{\text{tip}})/t \text{ lb}_f \text{ average impact force}$$

This force is not constant over the time period but is more like the graph of Figure 7.3.3. Since the time (t) is the same, the areas are made equal.

$$F_{\text{peak}} = 2*(W_{\text{ball}}/g)*(V_{\text{tip}})/t \text{ lb}_f \text{ peak impact force}$$

$$V_{\text{tip}} = \pi*D*\text{rpm}/60 \text{ in./s}$$

The liquid ball ingested or slug will be defined as follows:

$$W_{\text{ball}} = (\rho_{\text{liquid}})*(\pi/6)*(d)^3 \text{ lb}_f$$

The time (t) to pass through or flatten (d):

$$t = d/V_{\text{tip}} \text{ s}$$

The peak impact torque on the shaft:

$$T_{\text{impact}} = F_{\text{peak}}*D/2$$

This is a very simplistic analysis as all the energy of the impact doesn't go into developing this force. Also since the outflow is through an eye of the impeller, it's not really representative of what is shown. It does show that a liquid slug can cause the impact force (F_{peak}) to be large at high speeds. For comparison purposes, consider a 1/8-in. drop of rain hitting a car windshield at 60 mph and also at an impeller tip speed of 600 mph, which is around sonic velocity. The forces developed are 2 and 170 lb$_f$, respectively.

Consider the coupling stresses listed in Table 7.3.1 for various amounts of liquid ingestion (W_{ball}) with impeller $D = 20$ in., rpm $= 12{,}000$, and a shaft torsional mean stress $= 18{,}000$ lb/in.^{2}.

Table 7.3.1 Impact torque and stress on compressor shaft coupling.

Condition	d (in.)	T_{impact} (in-lb)	$T_{impact} + T_{mean}$ (in-lb)	Coupling torsional stress (lb/in.^{2})
No liquid	0	0	31,000	2,500
Liquid droplet	1/4	9,600	41,000	3,300
Liquid	1/2	38,600	70,000	5,600
Liquid	3/4	86,700	118,000	9,200
Liquid	1	154,000	185,000	15,000

The coupling stresses listed in Table 7.3.1 are with a stress concentration factor of 2 and are the maximum impact torsional shear stress in the coupling including the mean torsional stress. Impact of a 1-in. diameter ball volume could cause a crack to develop in a highly stressed area. From this impact, an instantaneous coupling torsional shear failure or a short-term fatigue failure could occur, depending on the magnitude.

All this analysis is intended for is to alert the failure investigation team on an area of concern. A safety audit that includes the knock-out drums, demister screens, high-level cut-out logic, possible liquid build-up in the suction or discharge piping (i.e. roll-back), along with the process operation and line-up should be on the checklist before start-up, if no other cause is found. Debris going through the compressor would use a similar analysis using a different density and shaft torques could be eight times higher. Even small pieces will usually result in damage to the compressor.

7.3.1 Summary

Droplets of water can erode high speed impellers on centrifugal compressors, but slugs can cause catastrophic failures.

Reference

1 Sofronas, A. (2006). *Analytical Troubleshooting of Process Machinery and Pressure Vessels*, 273. Wiley.

7.4 Weld Failures in Vibrating Equipment

Vibrating equipment can be machines such as vibrating screens, conveyor pans, or a combination of both, designed to transport, sort, or dry materials. Various types are used in many processing industries. They typically vibrate at 500 cpm (cycles per minute) and accumulate vibratory cycles quickly. They can exceed the materials endurance limit in a couple of days. These conveyors contain many welds, and while welds may not be a concern with constantly applied stresses, they can be with cyclic stresses.

These types of conveyors are excellent weld fatigue testing machines even though they're not intended to be. When a weld is not designed for fatigue, it will probably fail. Welds contain unavoidable defects and with elevated cyclic stresses cracks can grow quickly from these defects.

I'm not a metallurgist but have analyzed many such failures as a consultant. Overstroking, broken springs, incorrect operating procedures on well-designed equipment, or poor attention to details on new designs were usually the causes such as those shown in Figure 7.4.1a–c. Some sources [1] say that 90% of failures in engineering components are due to high cycle fatigue.

Figure 7.4.1a shows a poorly designed 20-ft-long conveyor/dryer that had been recently installed. It had cracked at so many welds throughout the conveyor that a decision on whether to rebuild or scrap the unit was required.

Figure 7.4.1b shows a system that failed after a diverter gate became loose, causing high-impact stresses on the welds resulting in cracks. The meandering nature of the crack growth indicates a rather low cyclic stress. When the loud "clanking" was heard, the diverter gate was locked back into place, but a sizable crack had already started and continued to grow.

Figure 7.4.1c represents one of several welds on a new vibrating conveyor/dryer that had its rocker arm stabilizer struts misaligned during manufacture [2]. This resulted in large bending moments throughout the conveyor and the resulting weld failures.

Welds don't have the typical endurance limit that solid steel bars have. This is because the endurance limit of a piece of steel at $1*10^6$ cycles is based on developing a crack, not when a crack has already started. Once a crack is present, it can grow quickly.

Conveyor vibratory service generates $1*10^6$ cycles on a weld in a day or two, so weld defects grow when cyclic stresses are high enough.

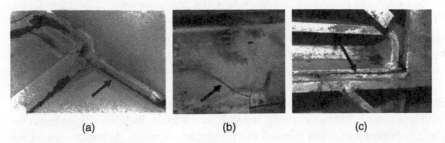

| (a) | (b) | (c) |

Figure 7.4.1 Dryer and conveyor weld failures.

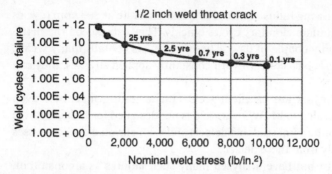

Figure 7.4.2 Crack growth in conveyor weld since new.

Fracture mechanics crack growth calculations [2, 3] are shown in Figure 7.4.2. This represents the approximate time for a small weld defect to grow through a good weld, with different cyclic nominal stresses opening and closing the crack.

Notice that a cyclic stress above $4,000\,lb/in.^2$ can result in a short life. A long life is expected below $4,000\,lb/in.^2$.

After analyzing many weld failures and knowing the approximate cycles to failure, it was noted that they all occurred at nominal weld stresses higher than $\pm5,000\,lb/in.^2$. This tends to agree with the threshold stress needed for a weld crack to grow [3].

For these reasons, it was decided to modify the conveyor purchase specification for clients. The specification stated that the nominal cyclic stresses at all welds in the structure should be at or less than $\pm4,000\,lb/in.^2$ verified by strain gage testing. Major equipment manufacturers said that this was possible by designing these welds for cyclic stresses and developing precision fixtures for the strut locations.

The statement "perform a better inspection" is not a good remedy because all welds have defects that can't be eliminated even if they are located.

For example, Figure 7.4.3a illustrates a "blind" plug weld that failed due to high bending stresses in the diaphragm plates it was attaching [2]. Since it can only be welded from one side, there is a gap where the crack started and the plate failed. The bending stresses in the design had to be reduced by thickening the plate and modifying the weld details. In Figure 7.4.3b, weld end discontinuities are removed by completely welding around and not allowing sharp-end weld details. This also reduces the stress on the weld.

Keeping the cyclic stresses low in well-designed welds is the most helpful method for avoiding weld fatigue failures. This is not always possible especially with an existing design. For these cases, there are many welding fatigue improvement techniques that can increase the fatigue life of welds. A brief review of the most used are shown.

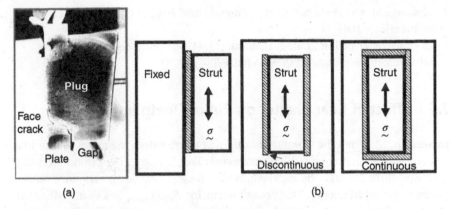

Figure 7.4.3 Plug weld failure and discontinuous weld.

- *Improve weld geometry*: Deposited welds usually don't have smooth transitions between the zones they join. Undercutting due to poor welding techniques can occur, which can greatly reduce weld fatigue life. Improving the weld geometry by grinding or by remeting areas such as the weld toe can help in lowering the high-stress concentrations due to the discontinuities.
- *Reduce residual stress*: High residual stresses can occur during the welding process due to the cooling of the molten weld pool and can result in high tensile stresses. While these are a constant tensile stresses and not cyclic, they can add to the cyclic tensile stress. These residual stresses can be lowered by methods such as shot peening or weld thermal relief methods.

Most manufacturing machine shops will have the tools necessary for the procedures described. They do require some training in the successful use of the methods.

The decision on whether to perform any of these techniques is of course based on the economics of the situation [4]. Some are quite time consuming and only marginally beneficial if done incorrectly.

7.4.1 Summary

Most welds will fail when their cyclic stress is above $\pm 4{,}000\,\text{lb/in.}^2$ unless special precautions are taken with the design.

References

1 Gurney, T.R. (1979). *Fatigue of Welded Structures*. Cambridge University Press.
2 Sofronas, A. (2006). *Analytical Troubleshooting of Process Machinery and Pressure Vessels, Including Real-World Case Studies*. Wiley.

3 Barsom, J.M. and Rolfe, S.T. (1987). *Fracture and Fatigue Control in Structures*, 2e, Prentice – Hall.

4 Kirkhope, K.J., Bell, R., Caron, L., Basu, R.I. (1996). Weld details fatigue life improvement techniques. *Rep. SSC-400, PB97-193031*.

7.5 Effect of Gear Chatter on Pinion Teeth Impact

In Section 7.1, when the vibratory torque was greater than the mean torque in a geared system, chatter or rattling was heard. This is due to the pinion vibrating back and forth through the backlash, which means gear tooth clearance, at the systems natural frequency. This doesn't occur for T_{vib}/T_{mean} less than 1 since the gear teeth don't separate to vibrate through the clearance. When T_{vib}/T_{mean} is less than 1, the load on a tooth may be high, but since the teeth stay in contact, there's no additional impact load. What happens when they separate and impact?

Consider a pinion gear interacting with the bull gear as shown in Figure 7.5.1.

A vibrating torque (T) is developed when an inertia (J) is displaced through $\pm\theta_{pinion}$ on an inertia (J) vibrating at $\pm$ (f_{cps}). A force F_{vib} due to this vibrating torque will be superimposed on the mean tooth force due to the mean torque. Both are additive, which means $T_{mean} + T_{vib}$, and will be imposed on a tooth face causing increased bending and contact stresses. An additional load F_{impact} occurs when $T_{vib}/T_{mean} > 1$. The force on the loaded tooth face is $F_{mean} + F_{vib} + F_{impact}$, and on the unload face that doesn't have a mean load is $F_{vib} + F_{impact}$ as it moves through the backlash (δ). This occurs on different teeth.

7.5.1 Summary

When the vibratory torque on gearing is greater than the mean torque, the gear teeth undergo impact loading on the front and normally unloaded back face of gear teeth. This may be heard as chatter and can cause high gear teeth contact and bending stresses.

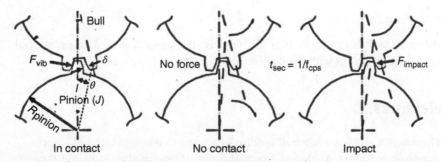

Figure 7.5.1 Chatter model through clearance.

7.6 Holzer Method for Calculating Torsional Multi-mass Systems

There are much more elegant methods for calculating the frequencies, mode shapes, and torques of a system than the Holzer method. The transfer matrix and finite element methods are two of the more popular methods used today. For a good understanding of what is happening without the use of matrices, the classic tabular Holzer method works well and is shown here in detail, so the readers can develop their own spreadsheet for troubleshooting purposes. Table 7.6.1 is an illustration of the tabulation method. This is for a seven mass system. For more than seven masses, just expand the table, and for less, reduce it.

The solution for each mode is by starting at $\omega = 0$, and increasing this value until F7 is near zero. When this happens, this is the natural frequency and mode shape and relative torque in the shaft due to the twist. By going up from this value to the next F7 near zero, you will have the second natural frequency, mode shape, and relative torque. Usually the author only uses this method for ω_1 and ω_2 for quick analysis and uses matrix solutions for higher orders due to accuracy concerns.

Remember $f_1 = 9.55*\omega_1$ cycles/min (cpm).

Here is what the table columns are showing:

Column B:

The torsional inertia value of the disk arbitrary amplitude $\theta = 1$ rad.

Column C:

The inertia torque for each disk.

Column D:

Angular amplitude rad at each disk first arbitrary at 1, and from this, the relative twist from H is subtracted.

Table 7.6.1 The Holzer tabulation method.

A	B	C	D	E	F	G	H
Item	J_i	$J_i\omega^2$	θ_i	$J_i\omega^2\theta_i$	$\Sigma J_i\omega^2\theta_i$	k_{ij}	$\Sigma J_i\omega^2\theta_i/k_{ij}$
1	0.166	4.63e5	1.00	4.63e5	4.63e5	5.7e6	0.08
2	0.33	9.21e5	0.92	8.46e5	1.31e6	5.7e6	0.23
3	0.33	9.21e5	0.69	6.35e5	1.94e6	5.7e6	0.34
4	0.33	9.21e5	0.35	3.20e5	2.26e6	5.7e6	0.40
5	0.33	9.21e5	−0.05	−4.56e4	2.22e6	5.7e6	0.39
6	1.84	5.14e6	−0.44	−2.25e6	−3.46e4	7.9e4	−0.44
7	20.8	5.81e7	0.00	3.45e4	−112		

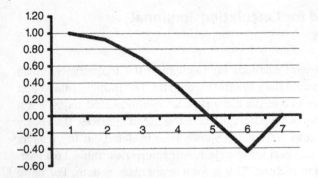

Figure 7.6.1 Graph of Table 7.6.1 column D.

Column E:
Inertia torque each disk at amplitude θ.
Column F:
Shaft torque beyond disk in question.
Column G:
Spring constants in.-lb/rad.
Column H:
Relative wind-up angle in each mass.
Assumed value $\omega_2 = 1{,}670.752$-seconds mode.

Notice that column F7 is almost zero and that ω_2 had to be taken to the third decimal place to get it there. That's why your spreadsheet will need an iterative procedure and why matrix solutions are used for higher frequencies.

Multiplying the relative values by the actual measured or calculated amplitude, as was discussed in Section 7.1, will result in the actual values (Figure 7.6.1).

7.6.1 Summary

The Holzer method is a relatively simple way to determine the vibratory natural frequencies and relative amplitudes of torsional multi-mass systems.

7.7 What to do When the Vibration Levels Increase on Large Gearboxes

Recently several large gear boxes that were in operation experienced increased vibration levels. The critical question asked to an engineer was whether the system had to be shut down. This is a major decision since in many cases, daily production losses may be substantial. Shutting down such a system, opening the gearbox, and finding no problems can be disastrous to one's career, especially if it was as simple as just bad instrumentation.

On critical gear boxes, these are the types of decisions that someone shouldn't make on their own. A team consisting of the owner's representative, the gearbox O.E.M., and the vibration analysis specialty company should present the supporting data to management with their recommendation. The owner's representative, who may be the site engineer, would lead the discussion.

The cases analyzed here are when the extruder or compressor system is still in operation. Once it is shut down, the gear box is usually inspected for damage. Unfortunately, this eliminates the opportunity to gather dynamic data to better define the problem. Is it a bearing failure or a gear failure? Using the proper spectrum analysis, time waveform techniques and observing the displacement, velocity, and acceleration data, the ability to show exactly what is failing in a geared system, while it is still operating, can be determined.

This is the approach many use when the on-line data is not complete enough to make an informed decision. When a continuous rise in acceleration or vibration amplitude with time is occurring, a shutdown would be a prudent decision because it might mean something is getting worse inside the gearbox. One piece of metal, like a gear tooth, getting into the mesh can make a 4-day repair into a 2-week repair. Again, this should be a team decision.

The most common causes of large gearbox failures are:

- Lubrication problems, which means water or particle contamination, the wrong lubrication, too high lubrication temperature, or no lubrication at all.
- Alignment problems due to bearings, couplings, or foundations can be external or internal to the gearbox.
- Load problems due to process changes or torsional vibration.

These causes can be determined by examining the gear teeth, bearings, or run-out when the unit is disassembled. Other methods are needed when it is still in operation.

Figure 7.7.1 shows a brief review of some typical problems noted when analyzing gear boxes on extruders, motor–gear–compressors, diesel engine–gearbox–propeller systems.

The time waveform shown in Figure 7.7.1b is the most informative curve for someone who only looks at one curve. Consider that there is a defect on one tooth of the gear as shown in Figure 7.7.1a. Every time that tooth is loaded, the force on the gear tooth changes and you get a spike every revolution the tooth goes through the mesh as in Figure 7.7.1b. When it is not in the mesh, the good teeth give a low vibration level when the normal gear load is applied to them. Each of the little lines represents one of the good gear teeth in the mesh. This would result in a vibration peak that occurs at shaft speed. Consider the case where all the teeth were damaged by the defect. All the small vertical lines would be bigger, even approaching the defective tooth amplitude. Since every tooth is damaged, it would be at the gear mesh frequency (GMF), which means the number of gear teeth times

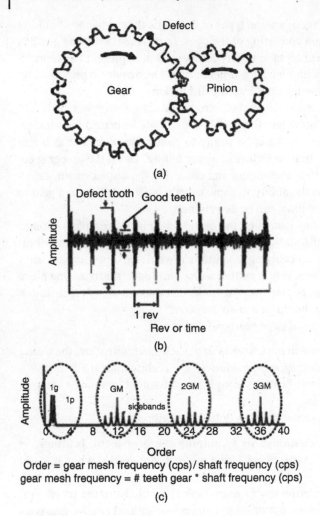

(a)

(b)

(c)

Order = gear mesh frequency (cps) / shaft frequency (cps)
gear mesh frequency = # teeth gear * shaft frequency (cps)

Figure 7.7.1 Time waveforms and typical FFT (fast Fourier transforms). (a) Gear defect, (b) time wave form analysis, (c) spectrum analysis FFT.

the gear rpm. So 55 teeth at shaft speed in cps of 10 cps is GMF = 10*55 = 550 cps. To simplify things, order is sometimes used, so this would be Order = GMF/shaft cps = 55 or the number of gear teeth.

The spectrum analysis or FFT shown in Figure 7.7.1c takes the time waveform and breaks it into pieces and averages them to basically summarize what is shown. The important thing to notice is that there can be a lot of information in this type of data and can be manipulated by knowledgeable specialists to diagnose the failure cause. Amplitudes above a certain level, which means velocity or acceleration, can indicate a gear tooth problem and will also indicate which gear. The side bands are also important as they indicate the modulation effect, which means load variation. In the case of one broken and the rest damaged teeth, Figure 7.7.1b would show the good teeth a little less than the broken tooth. The force would be modulating at 1× rpm. Alignment, broken teeth, worn teeth, cracked teeth, gear resonance, and much more can be diagnosed with the correct instrumentation and a knowledgeable person. Table 7.7.1 represents a summary of observations on gear boxes. Experience on extruders was that velocity readings above 0.5 in./s on the case near the bearing housing tended to indicate distress of a particular type. Readings less than 0.1 in./s were usually acceptable.

The idea that heavy bearing damage occurs in a gearbox is illustrated in Figure 7.7.2, which shows an unrecognizable spherical thrust bearing from a large extruder. The catastrophic damage was due to repeated impacts from trying to remove a cold plug of product in a polymer extruder [1]. There was no need

Table 7.7.1 Summary observed defect analysis.

Event	Source level (TWA)-time waveform, (FFT)-spectrum
Gear shaft misalignment	(FFT) 1 GMF but 2 GMF will dominate
Gear eccentricity	(FFT) 1 GMF and 3 GMF with higher sidebands. System natural frequency may be in the mix too
Gear broken tooth	(TWA) will show 1× rpm the shaft broken tooth is on (FFT) will indicate order
Gear heavy wear	(FFT) 3 GMF may grow usually in the first order range with 1× rpm sidebands. Many other harmonics also due to clearance and impacting. Gear ringing or system f_n may be in there too
Gear heavy loads	(TWA) (FFT) may show as an increase in the 1× GMF
Gear tooth chatter	(FFT) This can happen with torsional vibration when gear separation occurs $T_v/T_m > 1$. Each tooth impacts on the front and back side and may show up as 1 and 2× GMF. Numerous side bands and harmonics due to the tooth impacting
Gear growling, meaning a low frequency noise	(T) (FFT) Very low frequency usually because a defect on each gear comes into contact repeatedly every so often. Usually due to hunting tooth phenomena. Doesn't occur on gears that have prime number teeth, which means only divisible by themselves or 1. (For example, gear 53T is divisible only by 53 and 1, pinion 13T is divisible 13 and 1)
Bearing damage, gear case vibration, gear ringing, cavitation	(FFT peak G's) Usually shows as multiple harmonics of what the bearing defect passing frequency is. May show as high peak G's in the 1,000–4,000 cps range and not visible normal lower frequencies

Figure 7.7.2 Spherical thrust bearing impact damage.

for a vibration analysis. Faint striations indicate the motion of the impacts on the thrust bearing race.

In addition to the vibration analysis just discussed, there are things that can be done by the engineer when the unit is in operation, which are fairly straight forward:

- Take gearbox oil samples and send them out for oil particle analysis to a reputable laboratory. This may tell if it is a gear or bearing material. Make sure it's the correct lubrication and there is no water in it.
- Go out to the machine and take portable vibration readings and gear case temperature readings and verify that the vibration or temperatures are really as high as the recorded data indicates. This would be a sort of validation on the recorded data. Talk to the operators of the machine. They usually know the equipment well. Get baseline measurements from historical successful operation or similar operating machines if available.

When the gear box is opened for inspection, several things need to be done as outlined here.

Any gear box teardown has to be carefully planned with experienced personnel following the job. Don't rely on someone else if you are responsible for a successful start-up. Verify all the important steps:

1. Collect information on the gearbox and its history and interview the operators.
2. Visually examine gear teeth and bearing race patterns. Gears and bearings can indicate reason for gear teeth failures.
3. Document and photograph as much as possible, so it can be explained to others.
4. Torsional vibration load and gear calculations might be required to explain scuffing, pitting, yielding, wear, fretting, or tooth breakage.

5. Send critical specimens to a metallurgical laboratory.
6. Gather all the data and fit a failure cause that agrees with the observed data.
7. Recommend corrective actions to prevent future failures.
8. When new couplings are installed, remember to perform reverse indicator–type alignment verification with temperature corrections.
9. Thoroughly document important data.

Most of these items require a high skill level and experience on bearings and gear failures. Having the gear box representative on-site to assist has always been very helpful. The list can be much more extensive than this.

7.7.1 Summary

High vibration readings on operating geared units need to be compared to a baseline reading on a normal gear box, such as when new and verified by portable checks at the unit. Shutdown decisions require good data and should be a team decision. Continued increasing levels with time will justify a shutdown and inspection. Careful investigation with the correct instrumentation and experience can pinpoint the cause and the degree of the problem while the unit is still in operation.

Reference

1 Sofronas, A. (2006). *Analytical Troubleshooting of Process Machinery and Pressure Vessels*, 281. Wiley.

7.8 How Vibratory Torque Relates to Bearing Cap Vibration in a Gearbox

Taking vibration measurements on bearing caps is common and can be an indicator of gear distress in a gearbox. Direct measurement of the vibratory forces (F) would require torque or force measurements at the location which is usually not available. When an analysis and baseline data is available, the following method is useful.

It is usually thought that for torsional vibrations, the reaction torque is zero and nothing is felt externally. When gears are on the shaft, this isn't true [1] as can also be seen in Figure 7.8.1. They can now have an external reaction force (P).

In Figure 7.8.1, this might be the case with a shaft in torsional vibration.

If the reaction points (F) are near the bearing then they will displace (δ) with each vibratory force and be directly related to how flexible the point is. This means the attachment of the bearing to the gear case. This attachment can be determined

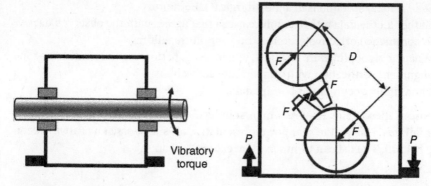

Figure 7.8.1 External reaction of vibratory forces.

from preliminary test and analytical data, which is basically the spring constant $k = F/\delta$. The vibratory torque is this force (F) times the gear pitch radius.

Consider that under ideal conditions, the gear reaction force vibration is calculated as $\pm F_{baseline}$ and displacement ($\pm \delta_{baseline}$) is measured at a given frequency, then under a new force ($\pm F_{new}$) and the same frequency.

$$\delta_{high} = \delta_{baseline}{}^* F_{new}/F_{baseline}$$

$$F_{new}/F_{baseline} = \delta_{high}/\delta_{baseline}$$

The peak velocity and acceleration are approximated using the following equations [1]:

$$V = 2^*\pi^*f_{cps}{}^*\delta_{high} \text{ in./s}$$

$$G's = \left(4^*\pi^{2*}f_{cps}^2{}^*\delta_{high}\right)/386 \text{ no units and called } G's$$

Since they all involve the same δ, V or G can also be used as long as the frequency is the same.

This is useful when performing analytical work such as torsional analysis as it approximates how much a bearing cap measurement (δ_{high}) might change with the new calculated vibratory forces ($\pm F_{new}$). The problem of course is that you must have good baseline value ($\delta_{baseline}$) and know the force ($F_{baseline}$) associated with this. With this amplitude, the following are for pure harmonic vibration.

Consider that the following information has been obtained from a torsional analysis of an extruder system, and the new gear force is required:

$$V_{baseline} = 0.2 \text{ in./s peak}, \quad V_{high} = 0.6 \text{ in./s peak}$$

$$F_{new}/F_{baseline} = V_{high}/V_{baseline} = 3.0$$

That's a significant increase and means something has made the force increase by a factor of 3.

This force can be compared to the results of the torsional analysis to help determine the cause of the increased vibrational force on the gear teeth. The frequency would be quite helpful in determining the cause.

7.8.1 Summary

Utilizing calculated data and bearing cap vibration measurements, an idea of what has caused an increase in a gear box vibration is possible.

Reference

1 Sofronas, A. (2012). *Case Histories in Vibration Analysis and Metal Fatigue*. Wiley.

7.9 Vibration of a Polymer Extruder Gearbox

Working with polymer extruders as a company engineer and later as a consultant to industry, gearbox vibration was always a concern. Extruders are used to process solids into pellet – like products by melting them and pushing them with a screw through a die. This extruder was a twin-screw unit in the 4,000-hp range, but there are also much larger ones.

This was an unusual extruder gearbox vibration problem as an identical extruder system was operating smoothly. All that had been done to this machine was to put a new gear-set in to be the same as the companion extruder. Figure 7.9.1 is an outline of the extruder system. The J_2 pinion and J_3 bull were changed out.

The concern was the high velocity and acceleration amplitudes recorded on the bearing cap housing of J_2 and J_3 at GMF. These were three times higher than the companion machine. The gear teeth were in excellent condition with 2 years of operation since new. As shown in Section 7.8, this suggests that something has increased the vibratory load on the gear teeth by a factor of 3. Screening studies are usually done when additional data is needed. In this case, a torsional analysis

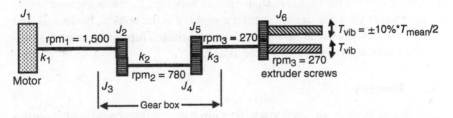

Figure 7.9.1 Extruder train.

Table 7.9.1 Extruder forced vibration resonance.

Excitation	T_v/T_m	Discussion
Motor side (J_2) with normal ±10% extruder excitation	0.2	No gear separation, so no chatter
Motor side (J_2) with ±50% extruder excitation	1.1	Gear separation, so chatter possible

was performed on the system to examine what might have changed. Torsional resonance probably isn't a concern since the gear teeth appear acceptable and the units have run successfully for many years.

Consider the extruder screw side excitation. Each screw under normal operation will experience ±10% or less of the extruder mean torque as vibration torque [1] at a frequency of twice screw speed, as shown in Figure 7.9.1. This is due to normal processing fluctuations. This level was also witnessed on a much larger extruder shaft with proprietary torque measurement installed. This is important since excessive screw vibratory torque may cause the gears to separate and chatter. Table 7.9.1 shows the motor side pinion load (J_2) with the normal 10% and excessive 50% vibratory torque on the extruder screw due to processing or wear. This is just forced vibration and not a resonance condition.

Table 7.9.1 indicates that with the normal extruder screw torque fluctuation, the motor side pinion (J_2) isn't of concern. However, if the fluctuation increased to ±50% on the extruder screws, possibly due to processing problems, then chattering of the pinion could be experienced due to gear teeth separation. This could result in an increase in the GMF. An increase in the GMF amplitude because of an increased tooth load could also occur when processing tougher polymers due to the increase in power required. The speed increase due to the gear change out was reported by the plant personnel as not being enough to increase the load.

It was also stated that there were no processing differences between the two machines. When there is the potential for several other causes, system mapping is usually recommended [2]. Mapping is a controlled testing procedure of the total system to identify the location of the highest vibration point and its frequency. For example, were the extruder shaft splines locking up or was there excessive extruder wear? Vibration at each point is measured in the vertical, horizontal, and axial planes. The critical locations are then investigated further by using measuring techniques such as discussed in Section 7.7.

7.9.1 Summary

The torsional analysis was used to see if certain abnormalities could contribute to a GMF high-amplitude vibration problem. In this instance, it shows that an increase in the extruder screw vibratory torque due to processing or wear above ±50% can cause the gear teeth to chatter, possibly at GMF. Also a higher toughness polymer would also increase the tooth load and, therefore, the GMF.

Additional data gathering in the form of vibration mapping and motor power versus vibration level were some of the recommendations made.

It is possible that the higher GMF may be due to slight internal misalignment and higher loads due to processing or extruder wear. Periodic gear inspection is, therefore, recommended as well as oil sampling and wear particle analysis.

References

1 Abeykoon, C. et al. (2009). Investigation of torque fluctuations in extrusion through monitoring of motor variables. *PPS-2009*, Cyprus (October 18–21 2009).
2 Sofronas, A. (2012). *Case Histories In Vibration Analysis and Metal Fatigue for the Practicing Engineer*, 33. Wiley.

7.10 Processing and Wear Load Increase in a Polymer Extruder

In Section 7.9, it was mentioned that a processing or wear increase could increase the GMF because of the increase in the gear load. It was stated that this might be some of the reason for the increase of the gearbox velocity vibration signal. The question is if such a load increase is reasonable when the throughput is said to have not changed?

Some idea of the extruder horsepower requirement is needed. The specific heat equation can help. We have used this equation for gases such as air and heating up or cooling down of solids, but it is also valid for the mass flow of solids.

The equation shows that knowing the throughput, specific heat of the material, and temperature of the product, the horsepower needed to accomplish this is known. Of course there are overall efficiencies (φ) involved.

$$Q_{BTU/h} = w_{lb/h}{}^*C_{BTU/lb\cdot°F}{}^*\Delta T$$

$$\Delta T = (T_{extrude} - T_{start})$$

$$HP = \varphi^*[(w_{lb/h}{}^*C_{BTU/lb\cdot°F})/2{,}544]^*\Delta T$$

HP ∞ Torque*RPM and (φ) is process efficiency.

The values are interrelated. For example, changing the RPM changes φ, ΔT, and $w_{lb/h}$.

Anything that alters the horsepower requirements such as increased wear due to rubbing or a different polymer processing condition will change the values. Based on this, the throughput (w) observed in production output might remain the same but be balanced out for increased horsepower. Possibly ΔT has increased.

The torque on the screw is also a function of the wear on the screw [1]. As the rubbing force increases, so does the horsepower without changing the output significantly.

7.10.1 Summary

It's always worthwhile to analyze the cause for a significant horsepower increase when evaluation GMF increased amplitudes in extruder systems.

Reference

1 Sofronas, A. (2006). *Analytical Troubleshooting of Process Machinery and Pressure Vessels*, 296. Wiley.

7.11 Vibration Charts Can Give Faulty Information

Determining if the vibration level is acceptable on an operating piece of equipment is usually done by reviewing charts that have been developed for that purpose. The chart shown in Figure 7.11.1 is one of many of this type. It is intended for use on external bearing cap measurements. From the chart, the displacement, velocity, acceleration, and the frequency are shown along with the acceptability at the particular level. These variables are uniquely related to each other for harmonic motion.

Superimposed on the chart are crosses which are based on failures. They are mostly gear box failures but also some centrifugal compressors. Gear teeth and bearing damage were usually the cause. Using such graphs to record failures is a good way to document them. Usually a note is written next to each one as a reminder of what had occurred.

One important observation is that some of the data, which means crosses on the chart, is in the fair range but have actually resulted in a failure. The chart is an average of data taken by various investigators on all types of machines. The failure points shown with the crosses were pretty much limited to geared systems. The chart wouldn't be appropriate for a proximity probe on a shaft, which is providing a direct reading of the shaft movement. The bearing housing measurements may be remote from the actual failure point where the forces are being generated and, thus, be attenuated.

Readings on a gearbox using spectrum and time waveform analysis techniques are discussed in Section 7.7, recording two identical gearbox with readings at the same locations on each unit. One recorded a GMF amplitude of 0.6 in./s at 50,000 cpm and the other 0.2 in./s at 50,000 cpm. The engineer whose historical

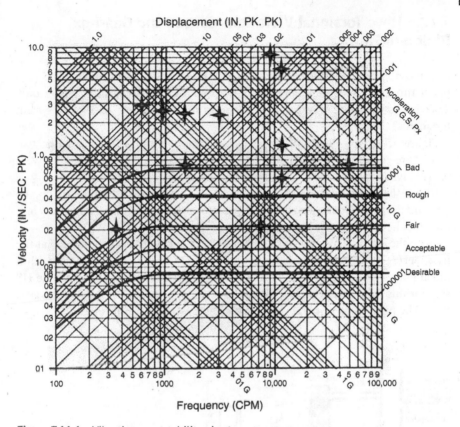

Figure 7.11.1 Vibration acceptability chart.

failures were marked on the chart would feel fairly confident that there were problems with the gearbox with the 0.6 in./s reading. This is especially true since the similar machine was reading three times less. That's because similar failures occurred in this range before on other equipment. A detailed analysis of the vibration data is warranted. A comparison of the loading of the two extruder system motors would be wise to ensure the increased vibration isn't just due to a higher loading.

7.11.1 Summary

Marking up charts with failure data you have witnessed is a good way to document past experiences. These types of failures may occur elsewhere.

7.12   Have Torsional Vibrations Caused the Gearbox Pinion to Fail?

A 3,100-hp diesel engine in a 1,100-ton fishing vessel was driving the propeller through a reduction gear unit. The pinion experienced pitting after only 2,000 hours in service. Was the problem with the gearbox design or system torsionals?

Figure 7.12.1 represents the mass elastic system outline. Only the first two modes are considered here for simplicity. The clutch and the engine modes were also analyzed in the analysis with the complete mass elastic system.

Notice that in this system, there are two excitation sources, one on the engine end and the other on the propeller end. They were considered independently and the propeller excitation is not analyzed here.

Since the engine has a viscous damper and flywheel, its mass is lumped at the flywheel (J_{flywheel}) for this analysis, since there no interest in the torsionals within the engine. Figure 7.12.2 shows the system analyzed referring to the engine HS side for this geared system. The springs and masses are modified using the gear ratio. The high speed side (HS) has a torsionally soft tuned quill shaft.

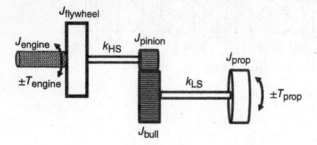

Figure 7.12.1   Mass elastic data for system.

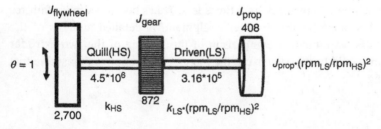

Figure 7.12.2   Mass elastic data to motor speed.

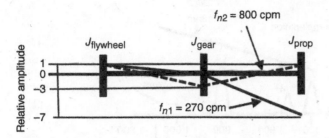

Figure 7.12.3 Frequency and mode shape.

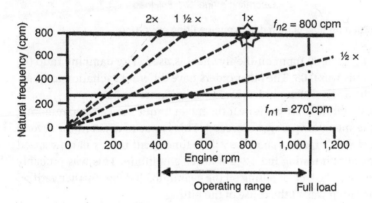

Figure 7.12.4 Campbell interference diagram.

Using the forced Holzer method of Section 7.1, the relative amplitudes of the first two modes are shown in Figure 7.12.3.

The second mode (f_{n2}) has been chosen to show the method since this mode excites the pinion. This is shown on the Campbell interference diagram given in Figure 7.12.4.

The vibratory torque on the pinion, which is often the cause of gear, shaft, and coupling problems, is always suspected. The pinion undergoes more cycles and motion than the bull gear. Excessive pitting on the front and back of the tooth face is an indicator that the $T_{vibratory}/T_{mean} > 1$ and tooth separation has occurred.

The excitation torque on the quill-pinion due to the engine firing torques is examined. There are many torque harmonics due to the engine firing impulses, and the critical ones that operate in the speed range all have to all be considered. Additional complications occur because the engine speed is variable. This is what makes analysis quite complex when internal combustion engines are the drivers.

For a sensitivity analysis of the (1e) order, which means the engine is exciting the system one time each revolution, excitation torque is determined from the firing order, bank angle, and pressure curve harmonics. For the (1e) order,

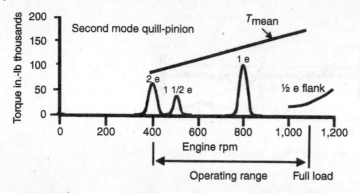

Figure 7.12.5 Torque on quill-pinion due to engine firing excitation.

$T_{flywheel} = 0.03*T_{mean}$ at 800 cpm on the flywheel is used. The damping ratio (ζ) used for the analysis was 0.05. The other orders have different excitation torques.

The result of the analysis for 4 orders is shown in Figure 7.12.5.

The (1e) critical vibratory torque is near the mean torque, and gear hammering could occur if the mean torque was exceeded. This is not of concern for several reasons. First, the ship's records show very little time spent in this narrow speed range. Also, torsiograph testing indicated a lower amplitude. This was probably due to gear damping. The lack of pitting on the gear unloaded face further verified that torsional vibration wasn't the cause of the pitting.

The pitting was eventually determined to be due to the gearbox pinion float being restricted. This caused one-half of the double helix or divided herringbone gear to take more of the load which resulted in the pitting due to a high contact stress.

The engine manufacturer was responsible for tuning the engine–gear–propeller for an acceptable torsional system. The gearbox manufacturer handled the warranty claim since it was a manufacturing error that caused the gear distress.

7.12.1 Summary

Systems driven with an internal combustion engine can be quite complex when performing a torsional vibration analysis. This is because of the many firing impulses of the engine that need to be considered. Careful tuning of such systems is required.

8

Examining the Human Body

To Question Is To Learn

8.1 What Causes Football Brain Injuries?

As a mechanical engineer who has routinely modeled machinery and equipment for over 50 years in industry, I try to use my abilities to show engineers the various areas that they can contribute to by using engineering techniques.

I recently read a book on concussions during the game of football [1]. As a grandparent and someone who enjoys watching the game, I asked myself the following questions:

- What damage does a heavy impact to the head do to the brain?
- Are these impacts cumulative, meaning do many small hits have the effect of one large hit?

Some approximate modeling and calculations can help answer the first question. The second question is addressed in Sections 8.2 and 8.3.

An important part of this analysis is understanding the impact velocity (V) of two simulated helmeted heads with the padding as shown in Figure 8.1.1.

Figure 8.1.1 illustrates two players' heads as spheres with the enclosed brain shown shaded. They are running toward each other with equal velocities (V) but in opposite directions and then collide. This illustration shows that when they collide the cross-hatched padding in their football helmets deformed (s) but the brain keeps moving in the same direction at the initial velocity (V) until it impacts the front of the skull.

A little discussion here:

- The brain is supported in a fluid so the assumption is it follows the velocity of the skull. This is like a person in an automobile without seatbelts. If the automobile stops slowly, the person stays in place. If the automobile hits a wall and stops

Unique Methods for Analyzing Failures and Catastrophic Events: A Practical Guide for Engineers, First Edition. Anthony Sofronas.

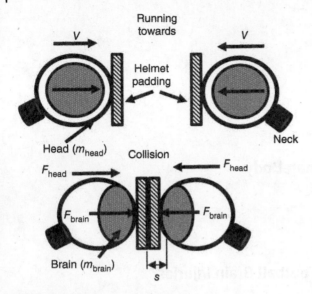

Figure 8.1.1 Head–brain impact.

suddenly, the person keeps moving at the velocity the automobile was initially and impacts the windshield. The brain mass impacts the skull.
- The brain and helmet will not see the same impact force because of the weight differences involve between the player's weight and the brain's weight. Just like a car crash, the car, which has more mass, sees more impact force than the occupant even though they decelerate at the same rate.

Each head sees an equal and opposite force (F_{head}). It comes to rest after the brain compresses and the force (F_{brain}) is developed. Only this frontal impact compression at a neck angle of 45° will be modeled to better understand what is happening. In this position, the head is like a battering ram producing a high force.

Let's consider a very simple model of the brain in the skull represented as the shaded cross section of a sphere as shown in Figure 8.1.2. For simplicity, only a frontal linear impact at a 45°-angle will be considered.

The velocities (V) of the two helmeted heads are coming toward each other and are equal but opposite. They each decelerate through the distance (δ_1) in time (Δt). The helmeted head masses (m_{head}) are equal. Using Newton's Second Law, the average force on the head at impact (F_{head}) over the impact time interval (Δt):

$$F_{head} = W_{head}{}^* V^2 / (2^* g^* \delta_1)$$

$\delta_1 = \delta_{cg} + s$ where δ_{cg} is the motion of the conforming brain mass

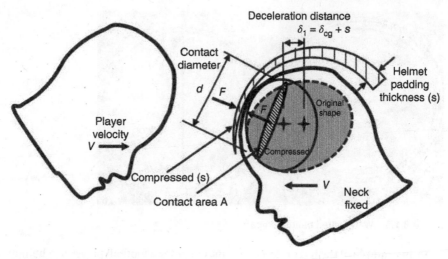

Figure 8.1.2 Model of head impact dynamics.

The peak force (F_{head}) on the head is twice the average force:

$$F_{head} = W_{head}{}^* V^2 / (g^* \delta_1)$$

The G's on the head and brain are approximately the same:

$$G = a/g = V^2 / (\delta_1{}^* g)$$

The force on the brain is:

$$F_{brain} = W_{brain}{}^* G$$

The average contact stress (σ_c) on the brain or pressure within the brain due to the impact:

$$\sigma_{brain} = F_{brain} / [(\pi d^2)/4]$$

The value (d) is taken as the cross-sectional diameter of the brain sphere as shown in Figure 8.1.2. A water-filled balloon representing the brain in the skull hitting a wall shows this diameter (d) shown in Figure 8.1.3. It also shows the balloon or brain expanding and conforming to the shape of the skull and filling the gap.

Some validation on this analytical model and the hydrostatic calculated brain stress is obtained from Nathum's experiments as reported in [2].

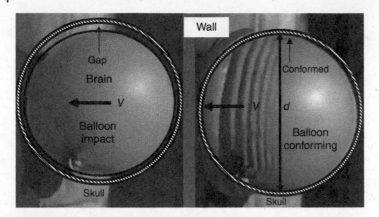

Figure 8.1.3 Water-filled balloon impact.

As an example of the use of the equations, consider a football player in a helmet hit by another player's helmet with each player being at a maximum impact speed of $V = 25$ ft/s.

$$G = V^2/(\delta_1{}^*g) = 103 \; G\text{'s}$$

$$F_{brain} = W_{brain}{}^*G = 310 \; \text{lb.}$$

$$\sigma_{brain} = F_{brain}/[(\pi d^2)/4]$$

$$\sigma_{brain} = 13.0 \; \text{lb/in.}^2$$

The force on the head:

$$F_{head} = W_{head}{}^*G = 2{,}100 \; \text{lb}$$

With forces and decelerations on the brain known how can this information be related to brain damage? Reference [3] shows moderate concussions occur above 100 G's or by the methods of this paper with a brain stress above 13 lb/in.^{2} as in Table 8.1.1.

Table 8.1.1 G deceleration limits.

G range	Stress on brain ($\sigma_{brain} = 0.126{}^*G$)	Effect on brain	Typical
Greater 300	38	Non-survivable	Unknown
250–300	32–38	Critical	Unknown
200–250	25–32	Severe	Unknown
150–200	19–25	Serious	Concussion, bleeding
100–150	13–19	Moderate	Concussion, severe car crash with seatbelt
50–100	6–13	Minor	Rocket sled, bruising
Less 50	Less 6	No injury	No bruising

The problem with this data is that it doesn't show the cumulative effect of many relatively light impacts. A method to do this is shown in Sections 8.2 and 8.3.

8.1.1 Summary

Helmets reduce head impacts from being life-threatening to a mild concussion. It has been shown that many concussions can lead to irreparable brain damage [1].

References

1 Laska, J.M. (2015). *Concussion*. Random House.
2 Nahum, A.M., Smith, R., and Ward, C.C. (1977). Intracranial pressure dynamics during head impact. In: *Proceedings of 21st Stapp Car Crash Conference*, SAE Paper No. 770922.
3 Rezaei, A., et al., Examination of Brain Injury Thresholds in Terms of the Severity of Head Motion and the Brain Stresses, *International Brain Injury Association*.

8.2 Life Assessment Diagrams

This method is used in Section 8.3. It is described in that section but to help clarify it here's a simple experiment used in seminars to explain it. A standard wire paper clip as shown in Figure 8.2.1 is used.

Of course, the brain is nothing like a metal paper clip but this example is used only to explain the cumulative fatigue and damage theory, which can apply to many areas.

A paper clip is bent in the manner shown until it fails and the number of cycles to failure is recorded. This is done for many paper clips, usually several per seminar participant.

Notice that this test is for an angle of 90° that is also a torsional stress.

This test could be done at any angle, for example, 45°, which would be a smaller torsional stress. With enough angles and tests, a graph of the cycles to failure (N_i) as shown in Figure 8.2.2 can be constructed.

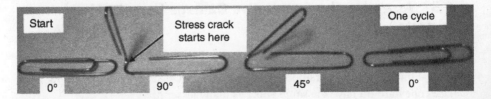

Figure 8.2.1 Bending a paper clip.

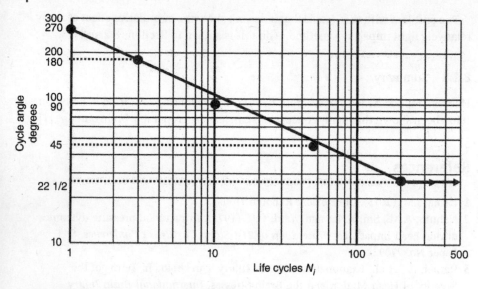

Figure 8.2.2 Cycles to failure of paper clip.

From this graph, the number of cycles to failure for a given bend angle, meaning torsional stress, can be determined. Notice that when the bend angle or stress is low, meaning below 22.5° for the paper clip used, no matter how many times it is bent it won't fail in fatigue. This is called the endurance limit of the paper clip material. Also because of a large angle or high stress, the paper clip fails after one cycle with a 270° bend. Only two points are needed to approximate this curve on log–log paper, failure after one cycle at 270° and after 200 cycles, the endurance limit at 22.5°. How this type graph is used to determine the cumulative effect meaning additive effect is discussed in Section 8.3.

8.2.1 Summary

Metals fail in fatigue when the stress is high enough to start a crack that grows with each cycle. Even if the cycling is stopped, the metal's fatigue life is shortened because the crack has already started to grow.

8.3 Assessing the Cumulative Damage Done by Head Impacts

In football, it's useful to have determined the stress on the brain from a one-time frontal impact, but what happens after many impacts at different severity levels? Does one hard hit have the same effect as many lesser hits?

Table 8.3.1 Game frontal impacts for a season.

Impact deceleration	Season frontal impacts
20 G's or less	164
50 G's	138
70 G's	31
90 G's	10
120 G's	3
150 G's	1
	Total season 347

Consider Table 8.3.1, the partial frontal impact data on a football player during a season which was obtained from telemetry using the data of Ref. [1]. The question is how do you know what the consequence of each impact is? Is one 150 G impact worse than ten 90 G impacts?

In engineering, there's a useful method to determine the remaining life of metal components that can fail due to metal fatigue and was shown in Section 8.2. These components undergo cyclic stressing over time. After a while, it will fail because it has been overstressed. A very small crack had developed and grew over time until the final breakage occurred. You can stop the bending back and forth at any time before it breaks. However, when you start bending again the metal "remembers", it has been stressed, because the crack is there and will grow. The life has been permanently reduced. In this case, it's because microscopic cracks have developed from the bending, actually torsion, and grow with each additional cycle. It fails when the crack goes through the wire. However, if the stress is low enough, the crack never develops and the bending can continue indefinitely. This is called the metals fatigue limit and is discussed in Section 8.2.

The brain may also have such a fatigue limit, but now this limit would be based on the degeneration of the brain and the cognitive abilities of the person. In the brain, it could be microscopic or macroscopic capillary or nerve damage impairing a portion of the brain. The same type of cumulative damage theory will be considered here. Failure is now defined as a degenerative-type brain disorder. The person is acting in a manner completely different than normal. When impacts are low, like being on a roller coaster (6 G's), no damage is probably done and can be continued indefinitely just like in metal fatigue. This lower limit has yet to be established but certainly walking with the brain seeing only 1 G is a lower limit. An additional limit is shown in Section 8.4.

Palmgren–Miner developed the linear damage or cumulative damage rule [2] and proposed a method to determine the remaining life of a metal component under cyclic stress.

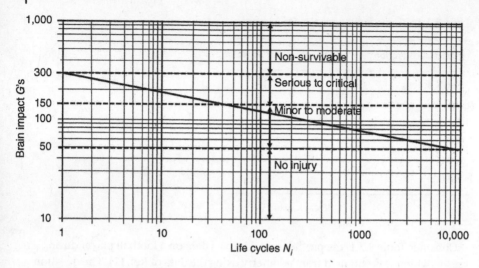

Figure 8.3.1 Fatigue and cumulative life curve.

This information will be utilized to see what the cumulative effect impacts on the brain have until permanent damage is done. A fatigue diagram as shown in Figure 8.3.1 that is also discussed in Section 8.2 is required for this type of analysis.

In developing this graph, only two points were available. To obtain such data, the individual's life impacts history of many players and their condition at the end of their careers would need to be documented. This would be a monumental task that hasn't been done yet. For this reason, an approach using existing published data to illustrate, the method is used here (Table 8.3.2).

The first point is when a fatality is said to be certain, which is over 300 G's and is shown as $N_i = 1$. Actually, this is not completely true as people have survived such

Table 8.3.2 Deceleration limits.

G range	Effect on brain	Typical
500	Skull fracture	Brain lesions
Greater 300	Non-survivable	Actually up to 500 G's survivable, depends on impact duration
250–300	Critical	Brain lesions
200–250	Severe	Concussion range
150–200	Serious	Concussion, bleeding
100–150	Moderate	Concussion, severe car crash with seatbelt
50–100	Minor	Rocket sled, bruising
Less 50	No Injury	No bruising

impacts and it depends on the duration of the impact. The second point is when there is no brain damage and is less than 50 G's at $N_i = 10,000$ and this number is suspected as the limit might be much lower. A line is drawn between these points. Only future experimental data will show if brain damage follows this trend.

The data used during a football player's season, Table 8.3.1 along with Figure 8.3.1, was used to test the Palmgren–Miner methodology to see if it appeared reasonable. The brain impact values, shown in Table 8.3.2, are calculated by the methods used earlier in Section 8.1 and the effect on the brain is from Ref. [3].

How do you combine various impacts to see how much life is used up before permanent damage is done to the brain?

The Palmgren–Miner linear damage or cumulative damage rule uses the following method:

$$\Sigma n_i / N_i = n_1/N_1 + n_2/N_2 + n_3/N_3 \ldots = 1$$

where

$n_1 =$ number of cycles of stress (impacts)
$N_1 =$ life cycles used up from Figure 8.3.1 graph.

Here's a simple example to address the earlier question if 1 hit of 150 G's is the same as 10 hits at 90 G's:

Consider the case where a player has been impacted 10 times with 90 G's on the brain for each impact:

$n_1 = 10$ cycles of 90 G's

Looking at Figure 8.3.1 at 90 G's, we read down to $N_1 = 400$;

so $10/400 = 1$

$0.025 = 1$ so 2.5%of the total life, which is 1, is used up.

Now for 1 hit at 150 G's,

$1/40 = 0.025$ or 2.5% so they have the same cumulative effect on the brain.

Notice that in Figure 8.3.1 for values lower than 50 G's, there's an unlimited life and those values wouldn't have to be considered.

Let's examine a player's seasonal career as shown in Table 8.3.1 if only the frontal impacts on a helmet were measured with telemetry. Since brain stress and G's are related, only G's will be used.

Under the impacts reported for this player for this season in Table 8.3.3, about 11% cumulative life of the brain has occurred. This seems high for a season since after 16 seasons of high school, college, and pro football all the brain life as calculated would be used up. However, the brain may be able to heal itself during the

Table 8.3.3 Fictitious players impacts season.

Measured G's	Number times G's occurred	N_i life from chart	n_i/N_i
150 G's	$n_1 = 1$	$N_1 = 40$	0.025
120 G's	$n_2 = 3$	$N_2 = 150$	0.020
90 G's	$n_3 = 10$	$N_3 = 400$	0.025
70 G's	$n_4 = 31$	$N_4 = 1,500$	0.021
50 G's	$n_5 = 138$	$N_5 = 10,000$	0.014
20 G's	$n_6 = 164$	$N_6 = \text{Infinite}$	0
		$\Sigma n_i/N_i =$	0.105

off-season. It does show that many small hits can have more of an effect than one heavy hit. For example, ten 90-G hits have the same effect as one 150-G hit.

The point where a concussion starts which is 150 G's in Figure 8.3.1 allows $N_i = 40$ cycles for a player's career (i.e. not a season) which seems quite high. Normally, players say they haven't seen that many even if they have played football through high school, college, and professionally [4]. However, many concussions may have been minor or unreported.

Saying that the season sample of 1 concussion from Table 8.3.1 is typical, a 16 season career would mean 16 concussion events of some magnitude occurred and would depend on the position played. Considering practice sessions maybe $N_i = 40$ is not unreasonable. With better data, it could be refined and used as a personal monitor of the brains condition of individual players.

8.3.1 Summary

Cumulative damage theory shows that many small impacts can have the same effect as a few large impacts. To know if this is true for the brain depends on how much the brain can heal itself, meaning reprogram itself during off-seasons. Helmets may help reduce the impact force like airbags in cars do, but the brain still has to decelerate and impact the skull. It does not seem likely that helmet designs alone can stop cumulative brain damage. Football is designed to be a collision sport that adds to the excitement. In my opinion, only better helmet designs along with rule changes could reduce the cumulative effect.

References

1 Rowson, S. and Duma, S. (2011). Development of the STAR Evaluation System for Football Helmets: Integrating Player Head Impact Exposure and Risk of Concussion. *Annals of Biomedical Engineering* 39 (8): 2130–2140.

2 Sofronas, A. (2006). *Analytical Troubleshooting of Process Machinery and Pressure Vessels*. Wiley.

3 Rezaei, A. et al., Examination of Brain Injury Thresholds in Terms of the Severity of Head Motion and the Brain Stresses, *International Brain Injury Association*.

4 Guskiewicz, K.M. et al. (2007). Recurrent Concussions and Risk of Depression In Retired Professional Football Players. *Medicine and Science In Sports and Exercise* 39 (6): 903–909.

8.4 What Happens When I Hit My Head and See Stars?

This can be answered by Section 8.3 on the cumulative effects of impacts. Why am I writing this section now? Simply because I just bumped my head and it reminded me. In Section 8.3, we saw that for someone playing football for about 12 years, meaning 4 years in college and 8 years professionally, say from 23 to 35 years old, a football player in certain positions can use up about 126% of the fatigue life of the brain using the unproven techniques discussed in Section 8.3.

I haven't played football but I have had several concussion events during baseball, basketball, motorcycle riding, or while working where I was knocked out for a couple of seconds. There were other cases where I saw stars and colors but didn't get knock out. How did this affect the fatigue life of my brain compared to that of a football player?

Table 8.4.1 is what I can recall from life experiences from about 15–50 years old. After that, I didn't do many hurtful things to myself.

This says that I've used up about 10% of the fatigue life of my brain according to the methods of Section 8.3. I was 75 years old when I wrote this and according to my wife and friends, my behavior hadn't changed over the years. I'm a mild-mannered person, still flying my aircraft, writing books, consulting and still remember things quite well. Happily the bumps and bangs I remember getting

Table 8.4.1 Cumulative damage to my brain.

Event type/number times	G's	n/N
Knocked out cold/2	150	2/40 = 0.05
Slightly knocked out/2	125	2/100 = 0.020
Saw stars/5	100	5/300 = 0.017
Heavy bumps/10	75	10/1000 = 0.01
Roller coaster type/10	50	10/10,000 = 0.001
		$\Sigma n/N \approx 0.10$

didn't seem to have affected my brain activity with only 10% of the fatigue life used up.

8.4.1 Summary

The normal bumps to the head that the average person gets probably don't affect their cognitive abilities much. Several G forces below 150 that occur quickly probably don't cause much long-term harm based on my experiences.

8.5 How Does the Body Keep Cool?

The skin is like a heat exchanger. The internal heat of the body due to the calorie burn and work done is conducted to the blood vessels just under the skin and by convection, radiation, and evaporation from the skin surface to the outside ambient environment.

Now the energy consumed by the body is called its metabolic rate ($Q_{metabolic}$). It can be thought of as heating up the body and can be 300 Btu/h when sleeping or 2,200 Btu/h for a bicyclist hard at work. The core temperature of the body is about 98.6 °F and the brain senses this and tries to control it with sweating. The brain can't control convection and radiation to keep it within a critical range as these depend on the outside environment.

Heat can be taken away from the skin to the outside environment by convection, radiation, and evaporation off of the skin.

The heat convected off of the skin surface is:

$$Q_{conv} = h^* A_{conv} (T_2 - T_3) \text{ Btu/h}$$

Radiation also removes heat to the environment:

$$Q_{rad} = 1.74^* 10^{-9*} A_{rad} \left[(460 + T_2)^4 - (460 + T_3)^4 \right] \text{ Btu/h}$$

The body starts to sweat when the skin temperature reaches 98.6 °F. The cooling evaporation from the skin has a normal maximum value of about:

$$Q_{evap} = 2,100 \text{ Btu/h with about one quart of sweat per hour liberated.}$$

The heat removed from the body must be enough to keep the body heat generated ($Q_{metabolic}$) constant or the core body temperature will increase and a heatstroke might occur.

Some nomenclature for the forthcoming equations is in Table 8.5.1.

Table 8.5.2 is the metabolic rates for various activities.

Consider the case of someone with light clothing bicycling on a $T_3 = 85$ °F day at 12 mph (18 fps). For this effort, $Q_{metabolic} \approx 2,200$ Btu/h.

For airflow off a surface (h) $\approx 1 + 0.225^* V$ where V is airflow in fps. With the core temperature (T_2) = 98.6 °F, the amount of heat dissipated is:

$$Q_{conv} = h^* A_{conv} (T_2 - T_3) = 5^* 10^* (98.6 - 85) = 675 \text{ Btu/h}$$

Table 8.5.1 Nomenclature used in equations.

T_1	Temperature of body internals (core) say equal to T_2	°F
T_2	Temperature of skin surface	°F
T_3	Temperature of ambient air	°F
h	Heat transfer coefficient off skin	Btu/h ft² °F
A_{conv} A_{rad}	Area of surface heat convected or radiated from exposed skin (about the same 10 ft²)	ft²
V	Air velocity over body	ft/s

Table 8.5.2 Metabolic rates for various activities.

Activity	Metabolic rate Btu/h
Sleeping	300
Relaxed	400
Walking (slowly)	700
Raking leaves	1,100
Walking (quickly)	1,200
Bicycling 12 mph	2,200
Running 9 mph	3,400

$$Q_{rad} = 1.74{*}10^{-9}{*}A_{rad} \left[(460 + T_2)^4 - (460 + T_3)^4\right]$$
$$= 1.74{*}10^{-9}{*}10{*}[(559)^4 - (545)^4] = 159 \text{ Btu/h}$$

The maximum sweat a body can produce is about 1 quart/h and evaporating off can remove about $Q_{evap} = 2,100$ Btu/h of heat.

The heat removed must equal $Q_{metabolic} = Q_{conv} + Q_{rad} + Q_{evap}$

$$Q_{evap} = Q_{metabolic} - (Q_{conv} + Q_{rad})$$
$$= 1,366 \text{ Btu/h or about } 1/2 \text{ quart of sweat.}$$

Since this Q_{evap} is less than 2,100 Btu/h meaning evaporation cooling is still possible.

With all of the heat removed, the core temperature should be stable as long as enough water is consumed to equal sweat lost. Going slower up a steep hill and this might not work since $Q_{metabolic}$ will increase and the airflow will decrease. With the core temperature too high, the rider might suffer a heatstroke.

When the outside temperature (T_3) is greater than 98.6 °F, then $Q_{conv} + Q_{rad}$ becomes negative and adds heat to the body instead of removing it and must be added to $Q_{metabolic}$. For example, if the outside temperature were 100 °F, then the evaporation rate would have to be $Q_{evap} = 2{,}373$ Btu/h to handle the new $Q_{metabolic} = 2{,}286$ Btu/h and there would be no convection or radiation cooling. Clearly, this won't work since the maximum evaporation rate of $Q_{evap} = 2{,}100$ Btu/h is exceeded and no additional sweat can be produced.

That's why hot temperatures can have a significant effect on the body. The heat transfer mechanism might not be adequate to handle the additional heat load. People who don't sweat are more susceptible to this. What happens is that the blood vessels dilate to get more cooling effect to the brain. This drops the blood pressure and results in dizziness, nausea, vomiting, weakness along with an increased heartbeat to try to cool things down. When not controlled, death could result.

While this engineering model didn't solve any major health issues, it did allow me to do a little research and to understand some of my body functions in engineering terminology rather than medical terms.

8.5.1 Summary

The blood vessels also transport and dissipate heat, like heat exchangers, from the body and the body controls evaporative sweat to keep the core temperature stable.

8.6 How Do Our Muscles Work?

I love to eat and while chewing on a steak, I realized that it took a lot of force to chew the meat. I wondered how all that force was developed and how much it was. Figure 8.6.1 illustrates the mechanics of chewing.

There is a difference between chewing on the rear molars (B_m) and biting on the front incisors (B_i). The largest force is on the molars because of the mechanical advantage as will be discussed. The jaw muscle (M) works by contracting and is represented by the spring. When it contracts, it develops a force (M). There are other muscles on the front of the jaw that aren't considered here. The pivot point (F) is when the lower jaw rotates and is also the fulcrum (F) shown in the free-body diagram. The lower biting and grinding molar tooth (B_m) and the front lower biting incisor teeth (B_i) are also shown. The reaction loads on the teeth can be determined by statics.

$$\Sigma M_F = 0$$

Doing this result in a mechanical advantage;

$$B_m = M^*a/b$$
$$B_i = M^*a/c$$

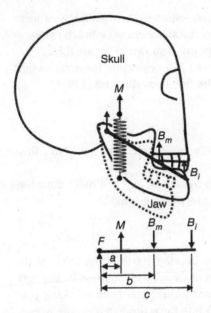

Figure 8.6.1 Jaw chewing and biting model.

So for a given muscle contraction (M), the molars (B_m) develop more force because b is smaller than c.

A pull scale test (ouch!) shows, I can bite with my front teeth with about $B_i = 30$ lb before my teeth start to hurt and that wasn't the maximum. Rough measurements show $a = 1$ in., $b = 1.5$ in., and $c = 5$ in. That would make the muscle contract with $M = 150$ lb and the molar bite $B_m = 100$ lb.

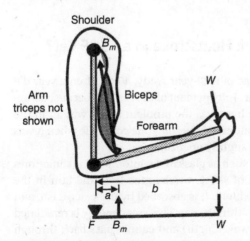

Figure 8.6.2 FBD of arm lifting.

Since the jaw has been investigated, the next question that immediately came to mind is how do my bicep muscles work? There doesn't seem to be much mechanical advantage or leverage at the attachment point as shown in Figure 8.6.2.

Summing the moments about the fulcrum (F) and neglecting the arms weight and angles, since they are almost vertical in the free-body diagram (FBD):

$$\Sigma M_F = 0$$
$$B_m = W^* b/a$$

Picking up a weight with one arm $W = 40$ lb with $a = 1.5$ in., $b = 12$ in., then $B_m = 320$ lb.

This is a large force for the bicep to develop by contracting and that's amazing. A weight lifter lifting 300 lb would have each bicep lifting 1,200 lb.

8.6.1 Summary

The jaw muscle contracts with 150 lb force to develop 100 lb biting force on the molars. From a mechanical engineering point of view, this would not be the optimum way to produce a high bite force. It does protect you from breaking your teeth due to the pain limiting, the maximum bite force produced on the nerves. Controlled bite tests in the literature indicate that this force could be 150 lb on the molars. The models geometry might be off or my teeth hurt too much to bite harder but it was close enough to understand the mechanism. The bicep shortening or contracting does this with a force eight times what it is lifting. Lift 40 lb and the bicep must develop 320-lb force. The triceps or gravity pulls the forearm down. The lack of mechanical advantage compensated for by the large force the bicep can produce when contracting. What causes muscles to contract to develop, this force is beyond my mechanical engineering understanding, however, most large body muscles work this way.

8.7 Why Do People Die from Heatstroke in a 75 °F Car?

A news report indicated on the average of a 20-year study, 40 children a year die because they were left in an automobile. Fifty percent of the time it's because someone forgot they were there. Excessive heat was the problem and it was all in the range of being 75–100 °F outside. I wondered how someone could die when it was only 75° outside and the window was slightly opened.

The buildup of heat across a barrier such as glass in an automobile is sometimes called the "greenhouse effect." Light of short wavelengths from the Sun in the visible range passes through the windshield. It is absorbed by the seats, dash, and carpets as well as the plastic interior within the car. This solar energy is reradiated at a longer wavelength (e.g. infrared wavelength) and cannot pass back through the windshield. It gets trapped in the car raising the temperature of the contents.

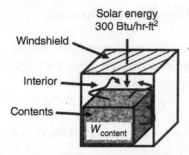

Figure 8.7.1 Automobile greenhouse effect.

Consider the simplistic model shown in Figure 8.7.1 to represent an automobile with its contents. The Sun's solar energy is passing through the windshield at the typical rate shown and no heat is lost through convection.

Solar energy q_{energy} is directed through the windshield and heats up the contents in the automobile. This energy is trapped in the car and using the specific heat equation:

$$q_{energy} = W_{contents}{}^* C^* \Delta T/t \quad BTU/h$$

$$\Delta T/t = q_{energy}/(W_{contents}{}^* C)$$

The solar radiation energy that comes through a windshield with an area of 12 ft^2 is:

$$q_{energy} = 300 \ BTU/h/ft^{2^*} 12 \ ft^2 = 3{,}600 \ BTU/h$$

$$W_{contents} = 200 \ lb, \ C = 0.4 \ BTU/lb°F \ \text{(average-specific heat of contents)}$$

$$\Delta T/t = 3{,}600/(200^*0.4) = 45°F/h$$

The temperature inside a closed car rises 45 °F each hour. Eventually, it reaches an equilibrium temperature usually within an hour or two.

This is important to know since if it's 75 °F outside, in one hour the inside of the car could be at 120 °F. The human body has a heatstroke at about 105 °F. Core temperature and children are even more susceptible. Even leaving the window open, a little won't bring the temperature down much because of the lack of airflow meaning convection currents.

8.7.1 Summary

The inside temperature of an automobile can rise 45 °F/h due to the greenhouse effect. Even on a 75-°F day, the interior can be at 120 °F which is in the heatstroke range. Leaving the window open, a little won't help much.

8.8 What Damage Can a Safety Airbag Do to a Human?

There is presently a concern with faulty airbags. Pieces of metal are being launched along with the airbag when it's deployed. Fatalities are said to have occurred. I wondered what the force would be from a flying metal piece.

Airbags deploy at a velocity of about 200 mph ($V_f = 293$ ft/s) in around 0.05 s and travel about 1 ft.

The airbag's uniform acceleration (a) can be computed from the velocities and distance moved (d) using the following formula:

$$a = \left(v_f^2 - v_i^2\right)/2d$$
$$a = (293^2 - 0^2)/(2^*1)$$
$$a = 42{,}900 \text{ ft/s}^2$$

The force exerted on an object is equal to the mass of the object times its acceleration ($F = ma$).

When G's $= a/32.2 = 1{,}333$ and $F = W^*G$'s

So if the piece of metal is a 0.75-in. cube, this calculates to $W = 0.0044$ lb

$$F = (0.0044)^*1{,}330 = 6 \text{ lb}$$

While the force seems low, this is the same velocity and weight of some low-power pellet guns so this can penetrate the skin and do damage depending on where it impacts and the sharpness of the metal's edges.

8.8.1 Summary

Airbags deploy at 200 mph and if any loose metal is present, it can cause serious injuries.

8.9 How Is Blood Pressure Measured?

We go into the doctor's office and the doctor puts a cuff around our arm, pumps it up, lets the air out, and reads a couple of numbers. What does it all mean and how can you determine the pressure inside an artery from outside the body? This is a question many of us might be curious about.

Back in Sir Isaac Newton's day when he was looking at pressure, he stuck a straw in a pipe and measured how high the liquid went up or in other words he used a manometer. In this way, he determined the pressure. He was also interested in the blood pressure in arteries so he designed a thin glass tube and stuck it in a vein and measured the pressure that way. This was done, rather painfully, for many years.

Luckily, the non-evasive painless method, we use today, was developed by S. von Basch (1837–1905) and is the one discussed here.

There are two pressures measured, the systolic or pressure when the heart is pumping and the diastolic which is the pressure when the heart is relaxed.

Here is how it's done. The cuff is put around the arm along with a stethoscope for listening to the blood flow. The doctor pumps up the cuff and stops the flow of blood. As the cuff pressure is slowly released, the first number read is the systolic and may read 120 mm of mercury (Hg) on the manometer. This is the maximum pressure in the artery. As the air is released further from the cuff, the doctor listens for no sound or the heart relaxed and is the diastolic and may read 70 mm Hg. So your blood pressure reading is given as 120/70 that is in the normal range.

For home use, we usually use the electronic versions that even pump themselves up, but they are basically doing the same thing.

8.9.1 Summary

The blood pressure monitor determines the pressure in an artery by measuring the pressure when the blood flow has stopped by an inflatable cuff. This higher pressure is called the systolic pressure and when the heart is relaxed called the diastolic pressure. It does this without having to measure the pressure directly in the artery. 120/70 mm Hg is considered in the normal range.

8.10   How Does the Heart Work?

Taking your blood pressure gives you three readings denoting your blood pressure. They are systolic that is the pressure in the arteries as the heart contracts or pressurizes, the diastolic as the heart relaxes and releases pressure, and the pulse. The pulse is the contraction and relaxation of the heart and veins and how many times this occurs in a minute. This provides important health data.

The heart does a lot of great things as it circulates the blood. It supplies oxygen and nourishment and also keeps the body cool and removes waste products. It does this using the blood and sends it to other organs, such as the lungs and kidneys to be cleaned or utilized by the body as needed.

As mentioned before, I know nothing about medicine but I do know my heart is important and would like to understand it a little better.

As an engineer, I understand pumps, piping, and fluid flow along with heat transfer. Maybe I can use some of this to learn about the heart in engineering terms.

Figure 8.10.1 shows a simplified heart as I see it in medical terms. Four valves are shown, which keep the flow moving in the correct direction. All chambers work together during a contraction and relaxation of the heart muscle. One side of the heart, the left side, has a thicker muscle because it pumps the hardest.

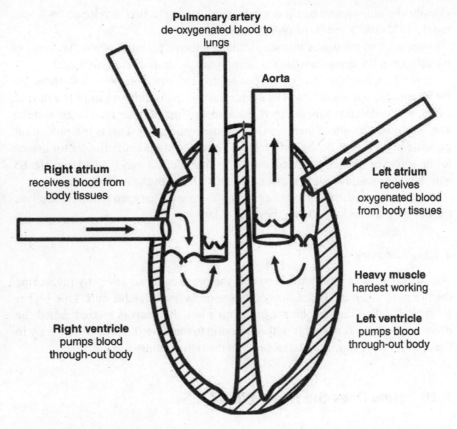

Figure 8.10.1 Medical heart.

Before building an engineering model, it's important to know the loop that blood is pumped.

- Arteries transport blood away from the heart
- Veins transport blood to the heart
- Capillaries transport blood from the arteries to the veins

Here's some interesting data on this wonderful system.

- The heart expands and contracts 60 times a minute or 3 billion times in a lifetime.
- The heart contains and pumps 6 quarts of blood a minute.
- The network of arteries, veins, and capillaries is over 60,000 mi long.
- Typical readings on the heart are 120 mm Hg systolic, 70 mm Hg diastolic

With a basic understanding of how the heart works, a very simple engineering model, with engineering terminology, can be sketched out in

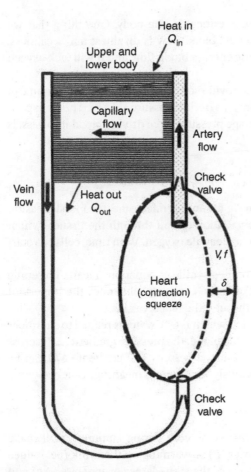

Figure 8.10.2 The engineering heart.

Figure 8.10.2. Using the data provided, some interesting results can be obtained. The four-chambered heart has been reduced to one chamber for simplification, since they all work together.

In the model when the heart contracts or is squeezed (i.e. the pump discharge), blood flows through the blood vessels meaning the arteries, capillaries, and veins (piping) and back to the heart (the pump suction). The pressure is determined by the resistance through the blood vessels (pipes) of equivalent radius (r). Check valves are in place to keep the flow going in one direction. The blood flow in the blood vessels also absorbs heat from the internal core of the body. It passes this heat to blood vessels just below the surface of the skin that cools it with the ambient outside air temperature (heat exchanger). Energy into the body is derived from calories (fuel) taken in by food and burned internally by energy expended.

The correct amount of blood flow is critical in the body. One thing that we can immediately feel with an inadequate blood flow is dizziness and weakness. The brain and body parts aren't getting enough blood flow and it will lack oxygen or become dehydrated.

The systolic and diastolic readings provide some valuable information and can be utilized in a quantity called the mean arterial pressure (MAP). This is experimental data that is the weighted average pressure force driving blood into vessels that serve the tissues.

MAP = (systolic + 2*diastolic)/3

For a normal BP of 120/70, MAP = 87 mm Hg

When this value falls below 60 mm Hg for an extended time, the blood pressure will not be high enough to ensure circulation to and through the tissues, which results in insufficient blood flow and inadequate oxygen. With time, cells can start to die.

An engineering model based on Poiseuille's equation (Jean Poiseuille 1797–1869) also called the Hagen–Poiseuille equation, illustrates the important variables affecting the flow of blood through the blood vessels.

It states that the blood flow (Q) of fluid which is 60% water is related to a number of variables such as the viscosity (η) of the blood, the pressure gradient (ΔP) across the length (L), and diameter (d) of the blood vessels. The gradient (ΔP) can be thought of as similar to (MAP) previously mentioned, meaning some weighted heart pressure.

Blood flow (Q) = $\Delta p/R$

One way to understand these terms is by considering sipping a milkshake through a straw. Make the straw shorter (L) or warming up the drink (i.e. reduce the viscosity [η]) or make the radius (r) of the straw larger or increase (Δp) and more flow will be taken in.

This analogy is true for blood flow also. Viscosity (η) of the blood is due to its makeup of platelets and can affect it. Intravenous therapy (IV) will reduce the viscosity by adding fluid to the veins. The blood flow is very sensitive to the blood vessel diameter (d) because it is raised to the fourth power. The blood vessels are usually flexible and pressure on them from skeletal muscles squeeze them open and closed to help pump the blood in addition to the heart.

For each of us, it is important to understanding what reduces blood flow. After all, this is what supplies oxygen to the tissues. So what can affect these variables and what does it mean?

- Advancing age and cholesterol buildup can reduce the average vein diameter (d) reducing blood flow. Reduce cholesterol by medicines or food restrictions.
- Advancing age and cholesterol buildup can reduce the pumping ability of the veins due to the loss in flexibility. Keep active.

- The viscosity can increase under some rare conditions such as liver disorders. This increases the viscosity and decreases the blood flow. Intravenous therapy (IV) will reduce the viscosity, increasing blood flow.
- Gaining about 10 lb of fat adds about 10% more veins to the body (L), reducing blood flow and makes the heart work harder to overcome the resistance.

Blood flow is complicated and when the heart's regular rhythm is irregular (i.e. arrhythmia), it may not pump enough blood when it is occurring (i.e. atrial fibrillation or AFib). Medicines may control this.

8.10.1 Summary

The heart acts like a positive displacement pump and the blood vessels are like pipes. They carry nutrients to organs and waste products from organs to be purified and recycled or disposed of. High blood pressure can overstress blood vessels and low blood pressure can lower cooling to the brain. Both are good reasons to keep your blood pressure controlled.

8.11 Restricting the Spread of a Virus

When I wrote this, we were in the midst of the coronavirus pandemic (i.e. COVID-19). It's called this because under the microscope, these types of virus strains have a halo or corona appearance. Catching the virus could result in fatal respiratory and various system infections and damage in humans. Worldwide, at this writing, it has infected 30,000,000 tested and killed 960,000. In the United States, it has infected 6,800,000 tested and killed 200,000. Borders have been closed, contact tracing of infected individuals is being performed and all but essential workers have been instructed to stay at home (i.e. self-isolation), keep at least 6 ft or more away from others (i.e. practice social distancing), wearing masks and washing hands frequently to stop from infecting others. This was a precaution since most people probably don't even know they have COVID-19.

This was especially concerning for me since my father had told me that when he was a young boy his four younger sisters had died in the Spanish Flu Pandemic of 1918 in Massachusetts/United States. He was saved because he was in Greece with his father, in a mountain village at the time. Over 1/3 of the world's population were infected in that pandemic and 50 million deaths occurred worldwide. In the United States, about 30 million recorded cases and 675,000 deaths had occurred. This happened even though at that time there was no significant air travel to help spread the virus.

Since I was isolated at home and am a curious engineer, I was interested in how fast COVID-19 could spread and how effective the implemented mitigation procedures had turned out to be.

Some of these were:

- Stop the in-flow of infected cases by shutting down entry into the United States.
- Contact trace individuals to determine who they may have infected and isolate them.
- Keep from being infected by implementing social distancing and self-isolation.

The virus seems to have an exponential growth rate. Figure 8.11.1 shows how the infection spreads from person #1 when each person can infect two others. This seems to progress as 1–2–4–8–16 and so on. This type of progression can be represented by the following infinite series:

$$2^0 + 2^1 + \cdots + 2^k = 2^{(k+1)} - 1 \text{ or in general;}$$

$$y = x^{(k+1)} - 1 \text{ or for a large number of cases it's approximately just } x^{(k+1)}$$

Let x be the number of people one person can infect, in this case, $x = 2$ which is the infection rate, k is number of time periods, in this case 4, and y is the total number of infected people, which calculates out to be $y = 31$.

This isn't usually the case and the number of people someone infects (x) varies based on the type of virus. It represents one of the primary unknowns of a new virus. Some viruses are much more contagious than others. For example, smallpox is highly infectious and x is between 5 and 7, meaning one person can infect 5–7 others. Fortunately, there is a vaccine that minimizes its spread. In the Spanish Flu Pandemic, x was probably around 1.4–2.8. For influenza, x is between 2 and 3 and for SARS (x) was about 2.75. H1N1 was the Swine Flu of 2009 and was a form of the Spanish Flu and was 1.4–1.6. It was lower because vaccines and antiviral

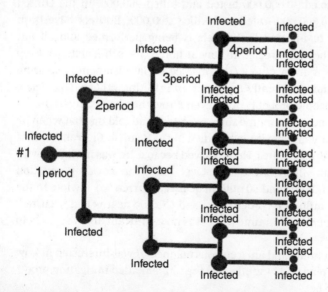

Figure 8.11.1 Growth of infection from person #1.

drugs made it less deadly. COVID-19 is estimated as being between 1.5 and 3.5. Only when this factor is less than 1 will new infections decline and die out? With a value of 1, the virus is alive but is stable and won't result in a pandemic. Assume for now that x for COVID-19 is the same as the influenza x meaning between 2 and 3 and the time period is 5 days between infections. For a month and a half the period, k would be 45 days/5 days = 9 periods and one person could infect the following number of people in this time period:

$$y = 3^{9+1} \approx 59{,}000 \text{ cases on the high side,}$$
$$y = 2^{9+1} \approx 1000 \text{ cases on the low side}$$

That's quite a difference and is an indication of the contagious nature of the virus. Being in error can result in over- or underestimating the number of cases. That's why new models on new viruses have to be continuously updated as new data becomes available. Inaccurate predictions at the beginning should be expected and the modelers aren't at fault. There just isn't enough good data to model with when in the early stages of a new virus.

This y or the total number infected is for only one infected person coming into a country and then closing the borders. The pool or group of infected is therefore limited. The group that came in or N_{group} is difficult to determine as it represents the number of people who had the infection at the start. Assume 1,000 people had entered instead of just one and then the borders were closed. The infected pool of people would be 1,000 times this number and assuming the lower infection rate of 2 then, the new cases in 1.5 months amounts to 1,000,000 cases.

$$y = (N_{group})^* x^{(k+1)} \approx 1{,}000^* 2^{(9+1)} \approx 1{,}000{,}000 \text{ cases}$$

There is a limit to the growth of infection in a fixed pool of people because people develop an immunity and recover and for other reasons. However, if nothing is done, the number of those infected can be enormous until the curve bends over and stops following the exponential equation. This is shown in Figure 8.11.2.

The peak of the curve is defined by the number of initially recorded infected people. In the United States, this peak seems to have been reached at 650,000 recorded cases after 1.5 months. Notice this is the recorded number and not the

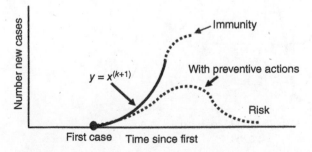

Figure 8.11.2 Limited exponential growth.

actual number. To know the actual number, the total population has to have been sample tested. After that, the new cases start to drop fairly quickly like a decreasing exponential series, similar to what was used to determine (y).

This simply means the exponential growth curve is flipped around and turned into a decay curve.

Of course, these are only the known cases because the whole US population, which is about 330 million, wasn't tested. However, if a path, meaning a line, in the diagram shown in Figure 8.11.1, can be broken, as it was, by removing and isolating people, they can't spread the infection to others. This results in the smaller curve with the 650,000 recorded case peak. While this is far from the actual number of people who have the virus, it does show that whatever the number it has peaked. This indicates that closing borders, contact tracing, social distancing, staying at home and eliminating situations where people are close can be very beneficial in reducing the number of persons infected.

Data was available [1] in September 2020 on the deaths due to COVID-19 and other hospital causes such as pneumonia in the United States. Figure 8.11.3 shows this. It seems to indicate a second wave has occurred.

8.11.1 Summary

Viruses spread exponentially. Restricting the flow of infectious persons into a country and then implementing guidelines to limit the spreading of those infected greatly helps.

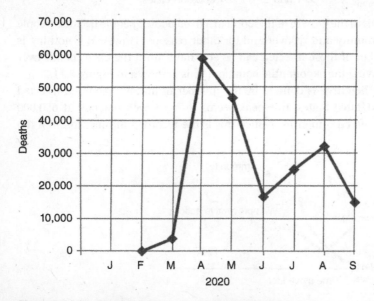

Figure 8.11.3 COVID-19 deaths in the United States.

The curves discussed during the pandemic are observation-type curves. As the infected cases are identified, they are plotted against time and a bell-shaped-type curve will eventually develop with enough data. It's not a statistical curve like a normal distribution curve, thus probabilities can't be assessed. Curves like this are developed for the country, each hot spot and each state.

With new virus strains like COVID-19, the number of people that one person can infect (x), meaning the infection rate is unknown as is the number of initially infected people (N_{group}) and the time between infections (k). The rate has to be projected or predicted from previous virus epidemics. This means that initially the predictions can be highly erroneous until enough cases are identified to define the infection rate, which will be well into the pandemic. Initially, only the extremes will be available from historical data, meaning the number of the most probable infected and least probable infected.

Another problem is that the reported and plotted cases are only those critical enough to be hospitalized with COVID-19 and other contributing illnesses or those that have been tested and documented. The vast majority will go undetected because not all of the population has been tested. This makes predictions of the future problematic.

This means that the decisions made, based on these type curves, will always be risky and this risk must be acknowledged. When to tell those in the population, they don't need to be concerned is a decision one person certainly wouldn't want to make without others. These are group decisions that all the experts in the medical and economics community have to agree on together. They need to realize that after the decision is made and any reoccurrence of the virus happens, they will be criticized by the nonparticipants. This is largely because those not contributing to the decision don't understand the fragility of the data used on which the decisions had to be made.

This all sounds quite familiar to an engineering consultant or problem solver. In engineering, this is similar to starting up critical equipment after a failure has occurred that caused fatalities. The decision to start up is a group decision and may have been based on a mathematical analysis and review of the failure data. Mitigating actions based on this information were probably implemented so the failure wouldn't repeat itself. There is always risk involved because the data and analysis methods can have errors. Eventually, the equipment has to be started up based on the group's best decision after analyzing all of the data. The company has to get back into production or it will go out of business and jobs lost. This sure sounds like what was occurring during the pandemic.

Reference

1 Center For Disease Control, Provisional Death Counts For COVID-19, cdc.gov/nchs/nvss/vsrr/covid19/index.htm, 2020.

8.12   Why Do Some Survive a Freefall Out of an Aircraft?

There are a few examples of people falling from great heights and surviving. Some have crashed through roofs like an airman who fell out of a B-17 during WW-2. A stewardess survived a free fall without a parachute after a bomb blast at 33,000 ft. She was cushioned by debris and landed in trees and snow and rolled down a hillside. She was severely injured.

Newton's Second Law that has been used a lot in this book can provide an explanation. As is sometimes said, "It's not the fall that kills someone it's the slowing down to a stop that does it."

In the following equation (F), lb_f is the force due on a body (W) lb_f traveling at (V) ft/s decelerating to zero in (s) ft:

$$F = (W/g)^* V^2 / (2^* s)$$

Assume a person weighing 150 lb has fallen out of an aircraft and has reached a terminal velocity of 176 ft/s (120 mph). How hard will the person impact earth?

$$F = 72,150/s \, lb$$

It seems it depends on the distance the body slows down in (s) ft. Breaking through roofs or trees will affect this distance (s). Obviously, landing flat on grass or water won't be effective and you will probably experience the full force if $s = 1$ ft.

Survival data on automobile crashes seem to indicate a blunt force much greater than 10,000 lb won't be survivable. Broken bones and ruptured vessels can occur. Landing on one's head or breaking one's neck would take much less force so the orientation of the landing stance is important.

A force of less than 10,000 lb would require (s) to be greater than 7 ft slow down distance. So possibly landing on your toes and slightly flexed legs and then tumbling like a gymnast does might help save your life.

8.12.1   Summary

Surviving a freefall from an aircraft or high place is usually not survivable unless, you are extremely lucky and also know a special technique.

9

Other Curious Catastrophic Failures Related to Earth

If It Isn't Calculation And Experimentation It's an Observation

9.1 Can an Asteroid Be Deflected from Hitting Earth?

After watching movies about asteroids heading toward Earth and how the world was saved, I decided to look at it from a mathematical point of view. I was interested in knowing if an impacter-type rocket hit an asteroid that was on a direct track toward Earth, would it deflect it and cause it to miss? So here's a very simple analysis to see if it can. Figure 9.1.1 is an illustration of the analytical model that will be developed.

For convenience, the rocket is shown as a sphere m_r at a velocity of V_r as it impacts an asteroid perpendicular to its trajectory shown as a sphere m_a which is at a velocity of V_a. Since the asteroid is moving at a velocity (V_a), this impact shifts its center by δ and sets it on a new trajectory. During the time (t), it takes for the rocket (sphere) to deform or disintegrate into the surface of the asteroid, the asteroid moves the distance (s) from the impact point and deflects in the direction shown. It is at a distance (D) from Earth. Will the distance (h) at which it passes Earth be enough to avoid a collision? Here's an analysis that will help answer the question.

Using Figure 9.1.1, the mass of an asteroid and of the rocket respectively are:

$$m_a = (4/3)*\pi*R_a{}^3*\rho_a/g$$

$$m_r = (4/3)*\pi*R_r{}^3*\rho_r/g$$

When the rocket impacts the asteroid, it is like a billiard ball hitting a stationary ball and sticking to it.

Unique Methods for Analyzing Failures and Catastrophic Events: A Practical Guide for Engineers,
First Edition. Anthony Sofronas.
© 2022 John Wiley & Sons, Inc. Published 2022 by John Wiley & Sons, Inc.

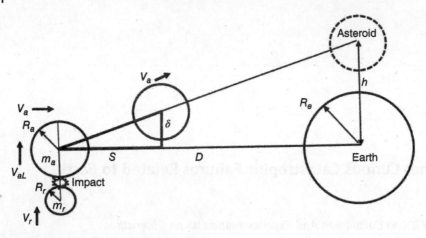

Figure 9.1.1 Asteroid deflection model.

The kinetic energy of the rocket impacter;

$$KE_r = (1/2)*m_r*V_r^2$$

Work done on asteroid $= m_a*g*\delta$

Equating and solving:

$$\delta = (1/2)*(m_r/(g*m_a))*V_r^2$$

Now let's assume the time that the impact takes place is the time it takes for the rocket to disintegrate into a crater. Conservatively data on meteorite craters say their depth can be 10 times the impacter diameter depending on the material:

$$t = 10*4*R_r/V_r$$

That means during this average time the asteroid moves toward Earth:

$$s = V_a*t$$

By similar triangles:

$$h = \delta*D/s$$

To clear Earth (h) must be greater than the radius of the asteroid plus the radius of Earth;

$$h/(R_e + R_a) > 1$$

For our calculations, the constants in Table 9.1.1 were used in the example: Table 9.1.2 shows the results.

An extinction event on Earth similar to what is thought to have helped cause dinosaurs to become extinct 66 million years ago can happen when an asteroid

Table 9.1.1 Constants used in analysis.

Constant	Value
Density of asteroid and rocket	$\rho_a = 150\,\text{lb/ft}^3$, $\rho_r = 500\,\text{lb/ft}^3$
Radius Earth	$R_e = 2 \times 10^{11}$ ft (4000 mi)
Radius asteroid	$R_a = 16{,}000$ ft (3 mi)
Radius rocket (impacter)	$R_r = 15$ ft
Velocity asteroid	$V_a = 7.3 \times 10^4$ ft/s (50,000 mph)
Velocity rocket	$V_r = 3.2 \times 10^4$ ft/s (22,000 mph)
Gravitational constant	$g = 32.2\,\text{ft/s}^2$

Table 9.1.2 Impact cases.

Distance	$h/(R_e + R_r) \gg 1$?	Arrival
$D = 10$ million miles distance from Earth, asteroid radius 3 mi , radius rocket 15 ft	Extinction event 0.02 ≪ 1 impact	8 d
Same as above but 1/4 mi asteroid radius	1.2 > 1 no impact	8 d

has a 3-mi radius. So under the conditions described, a rocket impact with such an asteroid 10 million miles from Earth will not provide enough deflection to miss Earth. Repeated impacts as it approaches could be successful. A 1/4-mi radius asteroid might be deflected from impacting Earth.

Of course, such a straight-line intercept is unrealistic since celestial bodies follow elliptical paths and the Earth is also moving away as shown in Figure 9.1.2. However, it does show that further the rocket impact on the asteroid is from Earth, the more chance there is for a successful deflection.

With the technology available today, a rocket may not get into position in time to impact and deflect the asteroid. Larger asteroids probably couldn't be deflected unless several heavier impacts were performed farther away. However, if they could be slowed by deflection, a miss is possible because of the orbital trajectory change.

9.1.1 Summary

It's possible to deflect an asteroid with a rocket if it is done far enough away from Earth or done repeatedly. Using celestial mechanics and trajectory calculations, closer distance misses with impacts might be possible because of the velocity changes.

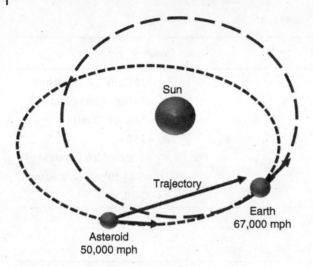

Figure 9.1.2 Earth moving away and trajectory.

9.2 What Size Crater Does a Large Asteroid Make When It Hits Earth?

There have been several asteroids passing between the Earth and the Moon lately. We were only told about them within a few hours before they passed by. An asteroid was said to be about 400 ft in diameter going around 57,000 mi/h and passed within 45,000 mi of Earth on 7/25/2019. Asteroid 2017 AG13 passed by 1/9/2017 and was said to be 100 ft in diameter at 52,000 mi/h and only 26,000 mi from Earth, which is way too close. Again no one told us until they were a few days away, as if we could have done anything to protect ourselves.

I was curious if some simple analytical model might tell me what size crater, the asteroid might make if it impacted Earth. I could then compare the analytical results with the Meteor Crater in Arizona and other impacts that have occurred on the Earth and the Moon. Meteor Crater was thought to have been caused by a 160-ft diameter piece of iron–nickel rock traveling at 42,000 mi/h. It produced a crater about 3/4 mi in diameter and 700-ft deep. Most of the meteorite was thought to have vaporized on impact since mining never found the meteorite in the crater.

Consider the very simple case of a rock with diameter (*d*), hitting the Earth vertically as shown in Figure 9.2.1. Why did I choose a vertical impact? Truthfully, it's because it's the simplest to analyze and understand. Figure 9.2.2 is a 1-in. diameter steel ball hitting sand at 10 ft/s and the pattern it makes as it penetrates the sand. Notice the flat bottom of the crater as it blows the sand sideways and then up to

Ejected and distributed W_S

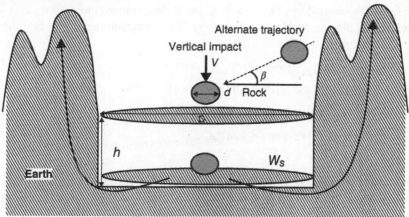

Figure 9.2.1   Impact crater model.

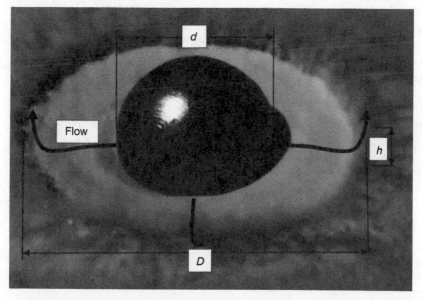

Figure 9.2.2   Steel ball impacting sand.

where it is then ejected out of the crater to be distributed on the rim or into the atmosphere. This will be the mechanism modeled.

The rock distributes the Earth (W_s) in the manner shown in Figures 9.2.1 and 9.2.2. The soil travels radially and then up and is ejected out of the crater until the energy is used up and the rock stops because this energy is dissipated. The energy is the work required to move the soil out of the crater. This is considered to be the weight of each grain of the displaced soil and the path it moves along or $(h + D/2)$ which is also the work done. The sum of all the radially moved grains is W_s.

Determining the weight of the rock (asteroid) and also soil removed when the rock stops moving:

Volume rock (V_{sphere}) and soil (V_{cylinder}) ft^3

Density rock ($\rho_r = 440$) and soil ($\rho_s = 200$) lb/ft^3

Weight of rock = $W_r = (4/3)\,(\pi*(d/2)^3)*\rho_r$

Weight of soil = $W_s = (\pi*D^2*h/4)*\rho_s$

KE = Work Done

This can be simplified since for these flat types of craters, there is a relationship $k = D/h$. For Meteor Crater in Arizona, $k = 5.5$ which was about the same for the steel ball and experiment shown in Figure 9.2.2 and in Table 9.2.1.

In actuality, probably only 1/4 of the energy goes into throwing the soil radially and the rest goes into moving the soil up and down along with compressing and vibrating the Earth and other energy vaporization losses. Let's call this fraction of the energy (α), where $\alpha = 1$ means 100% of the kinetic energy goes into moving the soil radially and $\alpha = 0.25$ only 25% of the energy moves the soil radially. It's value will be approximated from measurements of Meteor Crater and other craters.

The KE$_r$ when the rock impacts Earth;

$$KE_r = [(1/2)*W_r*V^2/g]*\alpha$$

The work done moving the soil after substituting $D = h*k$ and simplifying:

$$\text{Work Done} = W_s*(D/2 + h) = h^4*[\pi*k^2*\rho_s/8]*[k + 2]$$

Equating KE$_r$ = Work Done and the equations can be solved for the depth (h).

The crater diameter (D) = $h*k$

Table 9.2.1 Comparison of asteroid impacts and calculated values.

Case	Velocity (ft/s)	Rock Diameter (ft)	Actual crater diameter	Calc. crater diameter
Meteor crater	42,000	160	0.74 mi	1.1 mi
Extinction asteroid Chicxulub	40,000	10 mi	100 mi	87 mi
2019 unnamed 7/25/2019	80,000	400	No impact	3.2 mi
2017 BX 1/25/2017	25,000	30	No impact	0.3 mi
2017 AG13 1/9/2017	52,000	100	No impact	0.9 mi
1 in. steel ball/sand	10	0.08	4 in.	4 in.
Moon crater Potter & Kring	49,000	300	1.5 mi	2.0 mi

Here, $g = 32.2\,\text{ft/s}^2$, W_r lb, V ft/s, ρ_s lb/ft^3, ρ_r lb/ft^3, h ft, D ft, d ft, $k = 5.5$, $\alpha = 1/4$.

The model is very crude but it does provide some quantitative insight on the power of an asteroid.

Erosion is a factor but observes that the Moon with little erosion has flat dish-shaped craters. Earth craters were probably originally flat shaped too. Some reasons the analytical model may be in error are as follows:

- The rock may not impact the Earth vertically as shown in Figure 9.2.1 but at some angle β as shown as an alternate trajectory.
- The soil and meteorite densities may be different.
- The meteorite might explode or break up at some point during impact.
- The data provided on velocity and asteroid size may be incorrect.
- Energy is dissipated in other forms in addition to just removing the soil, such as heat and blast energy. Comparing with Meteor Crater, $\alpha = 1/4$ rather than the $\alpha = 1$ compares well with Meteor Crater Data so it was used in Table 9.2.1.
- The analytical model is too simple.

While they have been called asteroids, asteroids are when they are in space and they become known as meteors when in Earth's atmosphere. Meteorites are when they impact the Earth. Comets are something else and are not the iron-nickel core discussed here but are thought to be more like dirty ice balls in space.

9.2.1 Summary

Asteroids of even a 30-ft diameter that impact Earth can cause a crater with a 1,600-ft diameter. This is due to the high speed of impact. Most burn up in the atmosphere and never impact land but fall in the oceans since the Earth's surface is 70% ocean.

9.3 What Is an Earthquake?

We know that most earthquakes are the sudden jerking of the Earth; usually, because of the movement of the tectonic plates, the continents are on. This is a simple explanation of why they are violent. Figure 9.3.1 is a diagram that can help explain this.

Think of a fault line on the Earth as a big block of material like a rubber block. Why rubber? Rock can have a wide range of properties especially deep in the Earth. Consider the extremes from a liquid, meaning lava flows from volcanic action, to the hard brittle rocks we see on the surface. So they can be plastic meaning they don't go back to their original shape, to elastic which means they do. Like a piece of chalk you use on a blackboard, they can be brittle too. This means when a certain

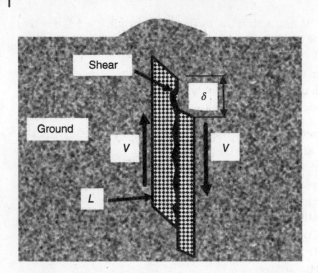

Figure 9.3.1 Movement of the Earth during an earthquake.

stress is reached they "snap" almost instantaneously. Now back to the rubber block of Figure 9.3.1.

One side of the block is being moved with velocity (V) in one direction an amount (δ) called slip in earthquake terminology and the other side the same amount in the opposite direction, thus putting the block in shear. The amount of shear can be defined as the strain in the material. Strain = δ/L and could be units of in. /in., ft/ft, mi/mi, or any other compatible length measurements.

When the strain reaches a limiting value, the rock will fail because its shear strength is exceeded. This is shown in Figure 9.3.2.

It would shear, meaning tear through the piece as shown in Figure 9.3.1. For example, if it was a hard rubber block, it might fail in shear if $\delta/L = 0.3$. In rock, this ratio is much smaller being about $\delta/L = 0.003$ and the shear strength (τ) is about 10,000 lb/in.^{2}. When it fails, it would also happen very rapidly.

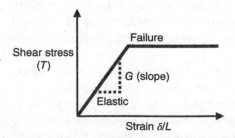

Figure 9.3.2 Stress–strain of a material.

The Earth or rock is not like rubber but it does fail with a sudden jerk when it is strained enough and follows a curve like Figure 9.3.2. The failures occur in a fraction of a second. The amount of energy stored in the Earth as it deforms is immense. For example, the stored energy can be explained as being similar to the potential energy (PE) stored in a spring, where k is the spring constant of the spring. It is the amount of force required to deflect the spring a given amount.

$$PE = \tfrac{1}{2}\,k*\delta^2$$

But k = applied force/deflection under the applied force:
= shear strength (τ)*area in shear $(A = L*L)/\delta$

Making the substitutions;

$$PE = \tfrac{1}{2}(\tau)*(L*L)\delta$$

Also at failure $\delta = \tau*L/G$, where $G = 2.9*10^6$ lb/in.^{2}. G is a property of the rock called the shear modulus of rock in its elastic range up to failure. It is shown in Figure 9.3.2.

Combining the equations is the energy when the piece of rock shears.

$$PE = (\tau^2)*(L^{2}*L)/(2G)$$

The fault line area in shear (A) for an earthquake will be assumed as $L = 1$ mi long by $L = 1$ mi deep. It would be several miles underground. All units are in the in. lb s system. It can then be converted to ft lb. The PE at failure would be $3.7*10^{14}$ ft lb. It may have been building up energy for many years and the δ at failure is $\delta = \tau*L/G = 18$ ft energy buildup before failure. As a comparison, the largest nuclear bomb ever detonated released $1.5*10^{17}$ ft lb of energy.

Luckily most earthquakes occur 5–50 mi underground. The shock waves are attenuated considerably by the time they reach the surface, which is discussed in Section 9.4.

Many small earthquakes along a fault line over time may be a good thing because they release some of the stored energy a little at a time. Turning it all into kinetic energy at once is what can cause catastrophic earthquakes.

This is all very crude but it does suggest the mechanism and the massive amount of stored energy involved.

9.3.1 Summary

Earthquakes can be from the sliding of one tectonic plate against another and the build-up of stored energy. Earthquakes occur when this energy is suddenly released with a slip or shear between the plates that can occur several miles underground.

9.4 Earthquakes Are so Strong Why Don't They Do More Damage?

This question on earthquakes naturally follows from those large energy release numbers in Section 9.3 on earthquakes. It was mentioned that earthquakes occur deep underground. As they travel to the surface, the shock waves are reduced. How much are they reduced?

The magnitude numbers of earthquakes are related to the Richter scale (Charles Richter, Physicist, 1900–1985), which show the effect of the earthquake on the surface and the distance from the earthquake epicenter which is on the surface. The hypocenter is the origin underground. This is shown in Figure 9.4.1.

From Section 9.3, the PE of the fault when it slips and shears was determined. To see the effect on the surface, consider that all the weight of the material (W_R) on top of the fault is lifted up δ_1 or the PE weight $= W_R*\delta_1$.

$W_R = L*L*d*\rho$ where L, d are in units of ft and

$\rho = 163 \text{ lb/ft}^3$ for the material.

For a cubic mile of material above the fault:

$= 5280*5280*5280*163 = 2.4*10^{13} \text{ lb}$

From Section 9.3, the PE for shear on area L^2 or a square mile of rock;

$PE = 3.7*10^{14} \text{ ft lb}$

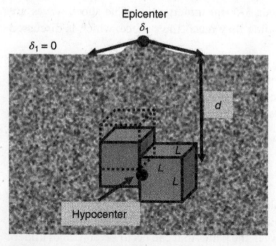

Figure 9.4.1 Earthquake underground.

Assuming only 1/4 of this energy is directed upward and the rest sideways and downward;

$$W_R * \delta_1 = (3.7 * 10^{14})/4 \text{ ft lb}$$

$$\delta_1 = [(3.7 * 10^{14})/4]/2.4 * 10^{13} = 3.8 \text{ ft}$$

So the epicenter will rise 3.8 ft from this energy located at its hypocenter 1 mi below the Earth. Now if the center is deeper than 1 mi, the weight of rock on top is heavier also so the amount of δ_1 will also be less as shown in Figure 9.4.2.

From Table 9.4.1 and Figure 9.4.2, a $1 * 10^{18}$ earthquake is similar to a Richter magnitude +8 and a $1 * 10^{14}$ a Richter magnitude 6.7.

There are many types of earthquake faults and all sorts of rocks and so this example is very specific. It does, however, show one way to determine the surface Earth movement from an earthquake deep underground. Its simplicity makes it quite problematic to someone well versed in earthquake behavior; however for me, I understand them a little better now.

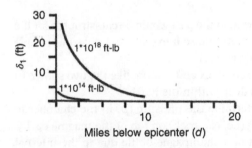

Figure 9.4.2   Ground motion verse depth.

Table 9.4.1   Earthquake magnitude and damage.

Richter	Energy (ft-lb)	Comment
2	$1 * 10^8$	Smallest noticeable
5	$1 * 10^{12}$	Small nuclear bomb
6.7	$1 * 10^{14}$	Northridge C. 1994
8	$1 * 10^{17}$	San Francisco, C. 1906 earthquake. Largest nuclear bomb (Tsar Bomba) ever detonated 55 Megaton
9.3	$1 * 10^{18}$	Sumatra, 2004 Boxer earthquake. 2 mi underwater 150 mi at sea
9.5	$1 * 10^{19}$	Chile 1960 most powerful in 100 years
13.0	$1 * 10^{21}$	Yucatan extinction event (asteroid) about 66 million years ago

9.4.1 Summary

Earthquakes occur deep underground and the effect when they reach the surface is much weaker. The Richter scale is a relative energy scale to determine the surface magnitude of the earthquake. A magnitude 2 is barely noticeable and a 9.5 is the largest felt in 100 years. A Richter magnitude 13 is what happened when an asteroid hit the Yucatan peninsula and eliminated most of the life on Earth about 66 million years ago.

9.5 Concerns on the Super-Volcano Under Yellowstone National Park

The favorite park of my families is Yellowstone. We love the beauty, the wildlife, and the geysers. There is so much to see. The geysers are caused by an underground volcano meaning a magma pool fed from deep within the Earth. It has erupted three times at 2 million, 1 million, and most recently 640,000 years ago and is quite active today as witnessed by Old Faithful, and all the other geysers, mud pots, and boiling lakes.

Eruptions like those that have occurred in the past would be catastrophic for life on the planet. I was wondering what would cause it to erupt so I did some simple calculations based on the model of Figure 9.5.1.

The magma chamber is much more complex and is more like the two shown in Figure 9.5.2 with both being fed from deep within the Earth.

All the simple model does is to determine the pressure (p) in the chamber to shear and lift out the cylindrical plug (i.e. cap rock) on top of the magma cavity. Basically, it's like a cork popping out of a champagne bottle due to the internal gas pressure, but certainly wouldn't be festive. It sounds simple but really is too complex for this type of analysis since so much is unknown. I'll give it a try anyway to learn some interesting things.

The volume of the cylindrical plug is shown as a dashed line and is a cylinder of diameter (D) and length (t). The weight when the material density is (ρ) is therefore:

$$W = \rho * \text{volume of cylinder} = \rho * (\pi * D^2 / 4) * t$$

The force trying to lift this weight is due to the cavity pressure (p) under it;

$$F_p = p * (\pi * D^2 / 4)$$

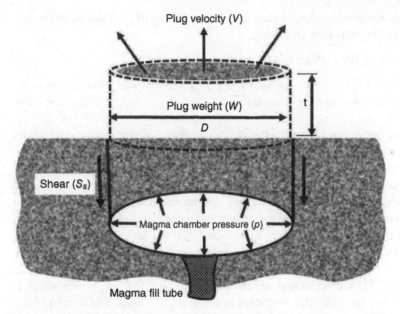

Figure 9.5.1 Model of molten cavity and earthen plug.

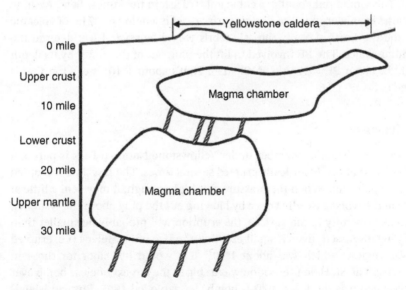

Figure 9.5.2 Yellowstone magma chambers.

The force keeping the plug-in place is the shear strength (s_s) of the rock material plus the weight (W) of the rock plug;

$$F_s = s_s{}^*\pi{}^*D{}^*t + \rho{}^*(\pi{}^*D^2/4){}^*t$$

When the ratio F_p/F_s is greater than 1, the plug will dislodge since the pressure force under the plug is greater than the shear strength and weight of the rock holding it in place.

The hard part is figuring out what D, t, s_s, ρ, and of course p should be. Units will be in the ft lb s system.

The chamber is thought to be 50-mi long and 12-mi wide. To be equivalent to the volume thought to have been dispersed 640,000 years ago, the equivalent circular area diameter (D) = 20 mi and the thickness (t) = 1 mi is used. The shear strength (s_s) of rock varies greatly but conservatively 2,000 lb/in.^{2} will be used [1]. The pressure (p) of 2,000 lb/in.^{2} [2] will be used. The density is (ρ) = 150 lb/ft^3. Of course, all of these numbers need to be converted to the lb ft s system.

Performing these calculations and the ratio comes out as F_p/F_s = 0.09 which is less than 1 so thankfully the plug stays in place, which it has. The weight of the plug has much more of a restraint effect than the shear stress. The internal pressure would need to reach 20,000 lb/in.^{2} for F_p/F_s = 1 and the plug theoretically would blow off. This would put about 315 cubic miles of ash in the atmosphere. Assume it distributes evenly around the Earth's surface and it would be 1/2 in. of volcanic ash around the Earth. Locally the ash debris would be several feet deep in the surrounding States. The PE involved to lift the plug out of the crater by its depth (t) would be about PE = $W^*h \approx 4{}^*10^{19}$ ft lb. That's about $1{}^*10^4$ mega-ton that is discussed in Section 6.6.

9.5.1 Summary

The pressure in a magma chamber under Yellowstone National Park is part of a super-volcano that has historically erupted several times. The last time it erupted was 640,000 years ago. When the pressure gets high enough, it may vent a little at a time or in the worst case all at once by blowing out the plug above the chamber.

Due to its proximity to the surface, the eruption will probably be smaller than the size plug discussed. Even a small eruption the size of St. Helens (i.e. erupted 1980, Washington) which was about $1{}^*10^{16}$ ft lb would be major for those in nearby states. The St. Helens eruption was as big as the largest nuclear bomb ever exploded which was the Russian Tsar bomb (i.e. exploded 1961, Russian island) or 50 Mega-tons.

References

1 Stowe, R.L. (1969). *Stress and Deformation Properties of Granite*, vol. C-69-1. Basalt and Limestone, U.S. Army Engineer Waterways Experiment Station.
2 Ibarguengoitia, M, A, 2011. Model of Volcanic Explosions of Popocatepetl, Doctoral Dissertation, Ludwig Maximillian University, Munich.

9.6 What Is a Tsunami and How Do They Form?

My wife had asked me this question while reading an article. She was satisfied when I said that's it's a water wave caused by an earthquake, volcano, or landslide under the ocean. However, I wanted to know more and since Sections 9.3 and 9.4 deal with earthquakes I thought I'd do some research.

Most of us have seen videos of tsunami waves coming on shore. Tsunami means "harbor wave" or "tidal wave" in Japanese. They are large, violent, and fast and destroy just about anything where they flow. Figure 9.6.1 is a sketch of a simple tsunami or tidal wave. In this case, I envision it as an uplift of the ocean floor due to an earthquake that displaces the water above it in a half-sphere bubble of sorts. The bubble in the form of waves moves out in all directions at velocity V.

Notice that there is a fault that has slipped up δ ft. When it does this, it displaces the water above it. This is like the analysis in Sections 9.3 and 9.4. Now instead of having solid earth on top of the fault where the density of rock is $\rho = 132\,\text{lb/ft}^3$, it now has the density of water or $\rho = 64\,\text{lb/ft}^3$. Consider that the velocity of the fault occurs almost instantaneously like in $t = 0.1\,\text{s}$. Then if the fault moves up $\delta = 75\,\text{ft}$, the wave might be moving at 500 mi/h. Such waves and speeds have been recorded. Smaller slips would be smaller and mover slower. The size of the wave increases as the ground slopes upward toward land and pushes the wave higher as it advances forward.

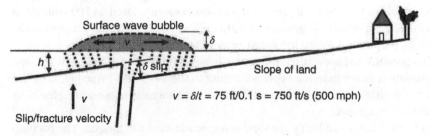

Figure 9.6.1 A tsunami.

Some authorities [1] using much more sophisticated methods suggest the following for a surface waves velocity based on the depth h(ft) of activity;

$$V = (32.2^*h)^{1/2} \text{ ft/s}$$

This means a disruption at a depth (h) of 2 mi ($h = 11,000$ ft) would be traveling at about 600 ft/s (400 mi/h).

9.6.1 Summary

A tsunami or tidal wave can be caused by an underwater uplift that displaces the water. It can race along at up to 500 mi/h and can be 100-ft high. The sloping land increases its height as it nears land.

Reference

1 Craig, W. et al. (2006). *Journal af Dynamics and Differential Equations* 18, 525–549.

9.7 What Is a Tornado?

We see tornados on TV and sometimes in real life. I remember as a kid on a farm seeing those dust devils break up bales of hay in the field and toss them in the air. I ran into one once to see what it was like. I got bounced around a lot and got some scratches from the debris in the little devil. I don't think it was fun because I never did it again.

When we see tornados, two things are usually obvious and those are the rotational speed and the base width, meaning diameter. The indicator on the strength of tornados is given as the Fujita scale with F5 being the strongest. Pressure differentials at the center at ground level have been measured as 100 millibar or (1.5 lb/in.^{2}). Atmospheric pressure is 1013 millibar. So if someone was inside the tornado, they would record the pressure as 913 millibar.

The pressure between streamlines gets lower with higher velocities when one streamline is faster than the other according to the Bernoulli Principle. Let's use this fact to determine the pressure differential in a tornado to get some idea of its potential for damage.

Figure 9.7.1 shows a bird's eye view of a tornado near the ground. The rotation in revolutions/min (rpm) can be determined from a distant observation, meaning debris rotating around. In addition, some idea of its width (D_1) might be guessed at.

In natural free vortex flow like a tornado, the velocity is inversely related to the distance to the center (r), meaning the velocity increases as you go into the tornado.

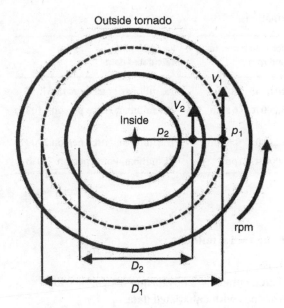

Figure 9.7.1 Simple geometry of a tornado.

This type of vortex is seen as water goes down a drain.

$$V_1 = \pi D_1 * \text{rpm}/60 \text{ ft/s}$$

$$V_2 = V_1/r$$

For simplicity, let's assume a linear velocity distribution where V_2 has some value near the center. For now, assume $V_2 = 2*V_1$.

From Bernoulli's energy equation between streamlines [1];

$$p_2/\rho_2 - p_1/\rho_1 = V_1^2/2g - V_2^2/2g$$

Simplifying:

$$\Delta p = -(3*\rho/2g)*V_1^2 \text{ lb/ft}^2$$

This is the pressure differential from the outside of the tornado to the center. Where the constants $g = 32.2 \text{ ft/s}^2$, $\rho = 0.08 \text{ lb/ft}^3$

Assume that from observing the tornado, rpm = 1.75, $D_1 = 1{,}000 \text{ ft}$

$$\Delta p = 31 \text{ lb/ft}^2 = 0.21 \text{ lb/in.}^2 = 15 \text{ millibar}$$

Atmospheric is about 1013 millibar so this just means it's 15 millibar lower inside or a reading taken inside would be 998 millibar. The flow would then be from the outside inward. The lower the pressure from the outside to the inside, the higher the inward flow or winds will be.

So how does this compare with the dust devil I walked into?

Table 9.7.1 Tornado data and calculations.

Size	Tornado diameter and rpm	Remarks and calculated data
Dust devil	10 ft, 45 rpm	1.0 millibar Force person 6 lb
Tuscaloosa, Alabama 2011, 190 mph EF4	8,000 ft, 0.6 rpm	117 millibar, Force person 1100 lb
Moore, OK 2013, EF5	10,000 ft, 0.6 rpm	182 millibar, Force person 1700 lb
Wray, Co., 2016 EF2	2000 ft, 2 rpm	81 millibar, Force person 750 lb

As I recall, $D_1 = 10$ ft, rpm = 45

$$\Delta p = 1.0 \text{ lb/ft}^2 = 0.009 \text{ lb/in.}^2 = 1.0 \text{ millibar}$$

This is probably like a strong breeze.

Force on person = $\Delta p*$ frontal area person = $0.009 \text{ lb/in.}^2 * 650 \text{ in.}^2 = 6 \text{ lb}_f$

Table 9.7.1 compares some tornadoes with calculated data.

Dust devils can get quite strong. On the Internet, I saw a dust devil pick up a child to a height of about 15 ft. The child survived with minor injuries. The rotation was about 300 rpm and the dust devil width was 5 ft. The calculated force is 105 lb$_f$. This at least shows that this was possible and wasn't "fake news" and that dust devils are dangerous. As a point of information, dust devils build from the ground upward and are due to the ground getting hotter than the air above. Tornados drop down from supercell clouds.

9.7.1 Summary

A tornado is a rotating mass of air with high rotational velocities and winds.

Reference

1 Sofronas, A. (2006). *Analytical Troubleshooting of Process Machinery and Pressure Vessels*, 142. Wiley.

9.8 Can a Tornado Really Lift a House?

I was watching a movie called Twister and it showed a house being lifted in the air about 50 ft and tossed several hundred feet and wondered if that was possible. A couple of quick calculations might clear things up. Winds in a tornado can

Table 9.8.1 Fujita scale for tornados.

F-scale number	Intensity	Wind speed (mph)
F0	Gale/branches off	40–72
F1	Moderate/mobile homes off foundations	72–112
F2	Significant/trees up rooted, roofs off	113–157
F3	Severe/roofs, some walls off, trees down	158–206
F4	Devastating/houses leveled, cars thrown	207–260
F5	Incredible tornado/cars thrown 1,000 ft	261–318

be greater than 260 mph so is that enough to move a small farmhouse? House movers say small houses weigh from 80,000 to 150,000 lb without foundations or furniture. Let's use 200,000 lb with furniture and the force required pull it off the foundation.

The wind side of a house might have an area of 30 ft by 25 ft or $A = 750\,\text{ft}^2$

The pressure on an area for a wind of velocity in 260 mi/h (mph) is:

$$p\,\text{lb/ft}^2 = 0.005*\text{mph}^2$$

$$F = p*A = 0.005*(260)^2*750 = 254{,}000\,\text{lb}$$

With the wind swirling and lifting on all sides, it certainly is possible for a strong tornado to lift up and toss a house before even tearing it up.

On the Fujita scale, Table 9.8.1 which is a measure of tornado strength also indicates this is possible.

9.8.1 Summary

It is certainly possible for even a relatively small tornado to lift and toss a small house before breaking it all up. It just depends how it's fastened to the ground.

9.9 Can Straw Penetrate a Tree in a Tornado?

A popular story is that during a tornado a piece of straw can achieve enough of a velocity that it can penetrate a tree trunk. Section 9.8 shows that on the Fujita scale 5 wind velocities of above 300 mph are possible in a tornado and could lift a house. At this velocity, it is doubtful there would be any trees left to penetrate.

When a piece of straw doesn't bend, it probably wedges into the trunk of the tree as a nail does with a hammer blow. The failure compressive stress (σ_c) of the wood is therefore important as the point pushes the fibers apart.

This calculation will determine the stress on a tree a piece of straw traveling at some velocity can develop.

The force is due to deceleration (a) of the straw (W_s) to a depth (h) into the tree:

$$F_s = W_s/g*a$$

$$a = V^2/(2*h)$$

The weight (W_s) of a 12 in. piece of straw, 1/8 in. diameter is $W_s = 0.0005$ lb:

Also let's assume a velocity of (V) = 350 mph (513 fps).

The compressive strength (σ_c) for ash is about 5,000 lb/in.^{2} and A_s is the straw cross-sectional area pushing the wood fibers apart.

The wedging force required to penetrate the tree pushing the fibers to the side is:

$$F_w = \sigma_c*A_s = \sigma_c*d*h \, (h/d)$$

The force available from the decelerating piece of straw;

$$F_s = [W_s/g*V^2/(2*h)]$$

Equating the equations for F_w and F_s and solving for the penetration depth (h):

$$h = [(W/(2*\sigma_c*g))*V^2]^{1/3}$$

According to this analysis, it's possible for the straw to penetrate an uncracked tree trunk to a depth (h) of 0.17 in. if it doesn't bend and break.

9.9.1 Summary

It seems possible that a piece of straw could penetrate the trunk of a tree if the straw is traveling at 350 mi/h and remains straight. Tests show that after a certain amount of penetration about 1/4 to 2 in., depending on the type straw used, the straw breaks up. A piece of piano wire will go through a 1-ft diameter palm tree [1].

Reference

1 Mythbusters Series, Episode 61, Deadly Straw, 2012.

9.10 What Is a Hurricane?

Since a tornado has been analyzed in Section 9.7, the same simple type of analysis model should be capable of analyzing a hurricane. One thing that I never really understood was when the weather reports said that an aircraft had flown into the hurricane and determined the pressure to be 950 millibar. What did that pressure mean? After analyzing the tornado model that became clear since atmospheric pressure was 1013 millibar. The difference between atmospheric is a Δp of 63 millibar which cause the winds to blow strongest at the eye wall of the storm (1013 millibar to 63 millibar = 950 millibar inside the hurricane). There is much better data for hurricanes than for tornados since they are routinely flown into and measured.

The same equations used in Section 9.7 will be used here only the tangential velocity of the wind speed at the eye wall is used. This is because the rotation or rpm of hurricanes can't easily be measured.

Using a spreadsheet of the equation of Section 9.7, assuming a 200-mi diameter and recorded wind speeds at the eye wall, Table 9.10.1 was developed.

Table 9.10.1 shows a good barometric comparison with actual measured hurricane data.

Obviously, much more is happening than the model explains and is quite different than complex actual theories suggest. The model does explain the general effect of the variables as the hurricane strengthens and that was all it was meant to do.

9.10.1 Summary

A hurricane is similar to a much bigger diameter tornado. Low-measured pressure readings like 980 millibar indicate higher winds within the hurricane. A reading of 1013 millibar is normal atmospheric pressure at sea level.

Table 9.10.1 Measured and calculated hurricane pressures.

Category	Velocity (mph)	Measured (millibar)	Calculated (millibar)	Memorable storms
1	85	988	984	Hermine, Fl 2016
2	100	972	973	Isabel, N.C., 2003
3	120	955	956	Katrina, LA, 2005
4	145	930	929	Galveston, TX, 1900
5	155>	919	918	Michael, FL, 2018

10

Strange Occurrences and Other Interesting Items

Any Analysis Can Always Be Used In Some Form Elsewhere

10.1   What in the Force of a Ship Hitting a Whale?

I read an article [1] on ships hitting endangered blue whales and the damage being done to the whale population moving through shipping channels. The idea was to give the whales a better chance by reducing the ship speed from 18 to 12 knots in the harbor shipping lanes. It was supposed to give the whales a chance to get out of the way and also cause less damage to them. The ship owners would be paid $2,500 to slow down. I was curious of the force on the whale due to this lower speed impact force.

These seagoing tankers are 1,000 ft long and weigh 100,000 tons or more. Compare that to a blue whale of 100 ft length and 200 ton in weight. The ship probably wouldn't feel a bump when it hits the whale but the whale sure would. Let's use some simple physics to see how much force the whale would be hit with. Figure 10.1.1 is my rendition of a whale being hit by the tanker. The scale of whale to ship is close to actual.

Some assumptions are that the whale when hit deforms through its width (s). Since the ship is so massive compared to the whale, it's like the whale ramming into a fix body or the whale has the velocity for the model, meaning that it's moving sideways toward the ship and the ship is not moving. The kinetic energy (KE) can be set equal to the work done (PE) and solved for the average impact force (F).

$$KE = 1/2\,(W/g)^{*}V^2$$

$$PE = F^{*}s$$

Unique Methods for Analyzing Failures and Catastrophic Events: A Practical Guide for Engineers, First Edition. Anthony Sofronas.

Figure 10.1.1 Whale hit by tanker.

With the whale's weight $W = 400{,}000\,\mathrm{lb_f}$ (200 ton), $s = 15\,\mathrm{ft}$ (whale width), $V = 30\,\mathrm{ft/s}$ (18 knots), and $g = 32.2\,\mathrm{ft/s^2}$,

$$F = 1/2\,(W/(s{*}g)){*}V^2 = 373{,}000\,\mathrm{lb_f}$$

As a comparison, that's like a person being hit in their midsection with something sharp equal to their weight at 30 ft/s that will do a lot of damage. Even reducing the speed to 12 knots would only drop the impact force to $F = 166{,}000\,\mathrm{lb_f}$ that is still too much. All the speed reduction will do is to give the whale a little more time to get out of the way.

10.1.1 Summary

A whale hit by the bow of a large ship will probably be killed. Reducing the speed will only give it a little more time to get out of the way. You would probably have to drop the speed to a couple of knots, where F would then be about a 2 ton force, to just push it out of the way.

Reference

1 Barboza, T. (2017). Ships to slow down off California to save whales and cut pollution. *Los Angeles Times*.

10.2 How Much Wind to Blow Over a Tree

Here's a simple way to determine how tall something is with just a square piece of paper and no calculations. We'll use it to determine the height of a tree but it can be used to determine the height of anything.

First fold a square piece of paper in half as shown in Figure 10.2.1. Walk away from the tree a distance S that is determined from holding the triangle as shown until you just see the top of the tree. The height of the tree is $S + H$. Say you walked away from the tree 50 paces and since a pace is about 3 ft, then $S = 150$. If you are 6 ft tall, then the tree height is 156 ft tall. Usually, you can neglect H.

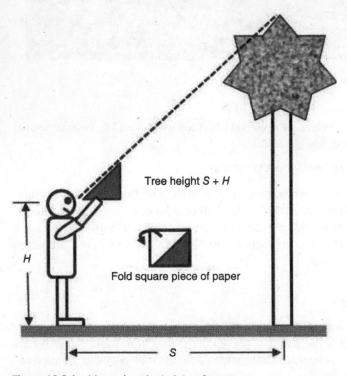

Tree height $S + H$

Fold square piece of paper

H

S

Figure 10.2.1 Measuring the height of a tree.

The key is that by folding the square paper, a 45° right triangle is formed so the two sides are of equal length. You add your height since you measure H above the ground.

Looking out a window at the trees in my backyard as a hurricane was entering, the Gulf is where this question How Much Wind To Blow Over A Tree came from. I could look to see damage done in previous hurricanes but I wanted to know if the trees in my yard could withstand a 120 mi/h (mph) wind. They seemed to have survived previous 60 mph gusts but they sure bent in the wind. Would my pine and oak trees survive? They had different heights, foliage, and root systems. This is the research and calculations done to answer the question. Usually, lightening was the culprit that took out the trees. They did this by splitting them or knocking limbs off of them.

Consider the simplistic tree with foliage and root system shown in Figure 10.2.2.

The tree has been simplified into its leafy foliage and its root systems both being balls with diameters D and d, respectively.

For the tree to fall over the wind moment, F_w*L must be greater than the restraining moment of the root system M.

The wind moment is easily calculated:

$$M_F = F_w*L = 0.005*A*mph^2*L = 0.005*(\pi D^2/4)*mph^2*L$$

The restraining moment (M) of the root will take a little research.

For many trees, the major part of the root system is equal to the drip-line diameter or $D \approx d$. So if you see where the leaves and branches end, that's the drip

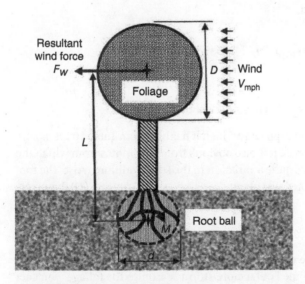

Figure 10.2.2 Tree equilibrium system.

Table 10.2.1 Resisting moment M for different trees.

Case	D (ft)	mph	L (ft)	M (ft lb) calculated
Oak Hurricane Katrina 2005	12	120	30	$0.25*10^6$
Oak Hurricane Bonnie 1998	12	85	50	$0.20*10^6$
Pine Hurricane Sandy 2012	15	115	40	$0.47*10^6$
Oak Hurricane Sandy 2012	15	115	30	$0.35*10^6$
Oak Tropical Arthur 2014	15	70	45	$0.20*10^6$
Pine Tropical Beryl 2014	15	65	45	$0.17*10^6$

line. Some trees like palms have a root that goes straight down and can usually withstand wind quite well.

The state of the soil the roots are in is important since rain-drenched soil will cause less resistance than hard soil.

Since M is unknown, a review of hurricanes and winds that have knocked over known size trees can be determined. From that, M can be calculated. A tree fell over when mph, D, and L were known so the resisting moment was M:

$$M = 0.005*(\pi D^2/4)*\text{mph}^2*L \text{ ft lb}.$$

Reviewing some historical data on uprooted trees on Table 10.2.1.

So to answer the question if the $L = 75$ ft pine tree, $D = 10$ ft in my backyard will topple with a 120 mph wind;

At 120 mph,

$$M_F = 0.005*(\pi D^2/4)*\text{mph}^2*L = 0.42*10^6 \text{ ft lb}$$

At 60 mph,

$$M_F = 0.12*10^6 \text{ ft lb which it has survived in the past.}$$

The pine tree will survive a 60-mph wind which it has but, from Table 10.2.1, is sure to go over at 120 mph. However, it is also obvious from the photos from which the data was taken that there was quite a variation in the soil conditions. Also, the root ball diameter (d) that uprooted was about the same as the drip-line (D) diameter.

10.2.1 Summary

Not many trees with foliage can survive 120 mph winds. It's a good idea to have trees trimmed, meaning thinned out, before storm season so the foliage won't act like a sail but will have room for the wind to blow through.

10.3   Why Do Objects Appear Smaller Than They Are?

Every few years we get to see a solar eclipse. The Moon between the Earth and the Sun perfectly blocks out the Sun to the observer on Earth so that only the perimeter of the Sun is visible. I was always curious as to why that was so. I know that I can block out a light bulb holding my thumb in front of my eye so it must be something like that. Actually, it is and it's just geometry as shown in Figure 10.3.1. Things look smaller when they are further away. This is how I reconciled it.

From the geometry, the apparent size h_1 of the object h_2 at a distance of d_2 is:

$$h_1 = h_2 {}^* d_1/d_2 \text{ or } 15{}^* h_2/d_2 \text{ with } h_1 \text{ and } d_1 \text{ in inches and } h_2 \text{ and}$$
$$d_2 \text{ same units like ft, in., or mi.}$$

The 15 in. is the focal distance you hold a frame from your eyes to look at the object from a distance that isn't blurred. You measure the size in the frame. I determined my distance by holding a book in front of me until the print just became clear. It came out as 15 in. I went outside and verified it on the size of my home. This 15-in. focal length might be different for you so it might be good to verify it.

So if you were 240 in. (d_2) away from and object 10 (h_2) inches tall, it would appear to be 0.625 in. (h_1) in the frame since $h_1 = 10{}^*15/240$.

There is another use for this equation. Hold your hand out and look at your thumbnail. My thumbnail looks about 1/2-in. wide at my eye. This means $h_1 = 1/2$ in. So if we know the height of an object h_2, we can determine its distance d_2 with our calibrated thumbnail.

$$d_2 = 30{}^* h_2$$

If I saw a car and I know it's about $h_2 = 5$-ft high and looked as big as my thumbnail it's about $d_2 = 150$ ft away.

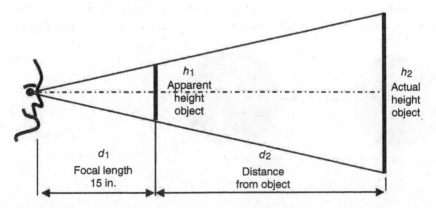

Figure 10.3.1   Geometry of size.

You can calibrate your thumbnail even further when d_2 and h_2 are the same units, for example:

1/4 thumbnail = $d_2 = 120^*h_2$

1/2 thumbnail = $d_2 = 60^*h_2$

1 thumbnail = $d_2 = 30^*h_2$

2 thumbnails = $d_2 = 15^*h_2$

3 thumbnails = $d_2 = 10^*h_2$

10.3.1 Summary

Because of the focal distance of your eyes and because of the geometry of distances, faraway objects look smaller to you.

10.4 Do We Feel a Force When Near Large Objects?

A discussion was going on about when you are standing near a large object, such as a pyramid, why your body feels different? Many people say that they do. Maybe it's just the majestic nature of size and how small one feels or possibly it's physical.

Newton's Gravitational Theory as discussed in Section 12.5 might provide an answer. At the least, it allows us to discuss this important law used in understanding the universe.

Figure 10.4.1 illustrates the law.

This theory states that every mass attracts every other mass by a force pulling the two masses together. Here, $F_{1,2}$ is the force between the mass, m_1 and m_2. The distance r is the distance between them. G is a gravitational constant, not to be

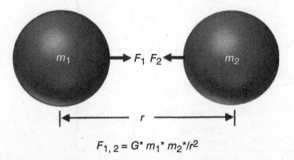

$$F_{1,2} = G^* m_1^* m_2^*/r^2$$

Figure 10.4.1 Newton's gravitational law.

confused with g (32.2 ft/s^2) the local Earth gravitational constant. G or Newton's constant $= 3.4*10^{-8}$ ft^4/lb s^4 with customary units to make it work.

Consider a pyramid weighs about $m_1 = 10*10^{12}/g$ and a person weight $m_2 = 200/g$, standing $r = 400$ ft from the center of the pyramid. The base of the pyramid is 800 ft.

$F_{1,2} = 0.4$ lb$_f$ attractive force

As a comparison, consider the force of Earth on a 200-lb person.

W_1 Earth $= 1.32*10^{25}$ lb$_f$, $m_1 = W_1/g$ lb s^2/ft

W_2 person $= 200$ lb$_f$, $m_2 = W_2/g$ lb s^2/ft

$r = 20.9*10^6$ ft from earth center

$g = 32.2$ ft/s^2

The force is trying to pull the person toward the center of the Earth:

$F_{1,2} = G*m_1*m_2/r^2 = 197$ lb$_f$

10.4.1 Summary

Any significant feeling you have toward the pyramid probably isn't because of the gravitational effect.

10.5 Why Does the Moon Sometimes Appear So Big on the Horizon?

I've always been astounded by how big our Moon looks when it's on the horizon. I know that sometimes the Moon's trajectory brings it closer to Earth but never to make it look several times bigger.

We know it can't be bigger and it isn't, it just appears so. One theory is that because there are objects to compare it to on the horizon, it just appears to be bigger. The trees, buildings, and other objects are what the brain compares it with. Consider the Ebbinghaus Illusion (Herman Ebbinghaus, Psychologist, 1850–1909) in Figure 10.5.1. Which of the dots in the center dot do you think is bigger?

Many would say the one on the right. Actually, they are both exactly the same size and illustrate how the mind can show you what it thinks you want to see. One object can make another object appear bigger.

Figure 10.5.1 The Ebbinghaus illusion.

10.5.1 Summary

The Moon stays the same size, and it's just the objects on the horizon that fool you into thinking it bigger. Put your thumbnail over it and then compare it later when it's high away from the horizon. It will still be about one-fourth size of your thumbnail. The diameter of the moon is 2,100 mi and it's always about 240,000 mi from Earth or from Section 10.3 about one-fourth of a thumbnail.

10.6 How Does an Air Conditioner Operate?

The south probably would never have had the population it has now if it wasn't for air conditioning. People would have chosen cooler climates. I know I wouldn't be living in Texas without it. Refrigeration came before air conditioners and they both work on the same principle. The thermodynamics of it all is a little too involved for this book but if you're interested an example is in [1]. For now, I'll just describe it in words and a sketch.

When a solid, liquid, or gas changes state, it absorbs a large amount of energy. For example, when a block of ice, a solid, melts into a liquid, meaning changes its state, and when the temperature changes to above freezing, it has absorbed heat energy. When water boils on the stove, a liquid, and turns into steam, a gas, it has absorbed energy to make this happen. So if we can find a liquid that turns into a gas, it will absorb energy, meaning cool things down.

Figure 10.6.1 represents a system to convert a liquid to a gas and it's the basis of air conditioning (A/C).

The unit outside the home contains the fan, condenser, and compressor. In the home are the evaporator coil, fan, and expansion orifice.

The system is pretty simple. The compressor compresses a hot gas and passes it through a cooling condenser where it cools it a bit and condenses into a liquid. The still hot liquid is then pressured through the orifice (i.e. a small hole) and is turned

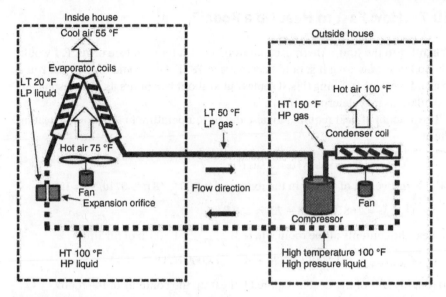

Figure 10.6.1 Home air conditioner system.

into a cold liquid and gas because of the pressure drop through the orifice. This cold mixture because of its change of state from a liquid to a gas absorbs heat in the evaporator coil where the blower blows air through it. This cools the air going into the rooms. In the evaporator coil, heat is heat taken out of it and it becomes a hot gas and goes to the compressor suction. The compressor compresses the gas to a high pressure and temperature and discharges it through the condenser. This condenses it back to a liquid and sends it to the expansion orifice where the process starts all over again.

Selecting a liquid with the correct pressure/temperature properties was key to making small efficient air conditioning units.

10.6.1 Summary

By developing a cycle that changes from a liquid, usually Freon, to a gas will allow heat to be absorbed from the hot room air and thus cool it. Many industrial refrigeration systems operate on the same principle but may use a different liquid instead of Freon. Pressures will be different too as will the expansion valve.

Reference

1 Sofronas, A. (2006). *Analytical Troubleshooting of Process Machinery and Pressure Vessels*, 187. Wiley.

10.7   How Fast to Heat Up a Room?

When I turn the house thermostat up from a room temperature of 70 °F, I would like to know how long it should take to reach 78 °F. One could always just run a test and measure it. Doing this, it comes out to about two hours. Is this reasonable? Calculations may determine if it is.

The amount of heat required to raise the air temperature in the room to heat up the air:

$$Q_{heatair} = W_{air}{}^*C_{air}{}^*(T_{end} - T_{start})$$

W_{air} is the weight of the air in the room and is 300 ft^2*8 ft*0.07 lb/ft^3 = 168 lb.

$$Q_{heatair} = 168{}^*0.2{}^*(78 - 70) = 269 \text{ BTU}$$

To heat the contents of the room where $W_{contents}$ = 1,500 lb of furniture

$$Q_{heatcontents} = 1,500{}^*1.0{}^*(78 - 70) = 12,000 \text{ BTU}$$

So roughly 98% of the heat is required to heat up the contents of the room.
Considering only the contents,

$$Q_{heat}/t_h = W_{contents}/t_h{}^*C_{contents}{}^*(T_{end} - T_{start})$$

The Q_{heat}/t_h term is the heat into the room from the furnace. For a 100,000-BTU/h furnace distributing through 15 vents, each room receives 6,700 BTU/h.

$$6,700 = 1,500/t_h{}^*1.0{}^*(78 - 70)$$

$$t_h = 1.8 \text{ hours}$$

10.7.1   Summary

The contents of a room require the most heat and it takes about two hours to raise the temperature of an insulated room 8 °F.

10.8   How Do I Size an Air Conditioner for a Garage?

While there are many tables available to determine the size air conditioner needed for a room, I was interested in A/C for my garage shop. Being analytical I thought I'd do some quick calculations to determine the BTU/h. required since that's what window units are measured in.

A useful equation is the specific heat equation and like in Section 10.7 only the garage contents are considered in the cooling down.

$$Q_{cool}/t_h = W_{contents}{}^*C_{contents}{}^*\Delta T/t \text{ BTU/h}$$

Now the weight of the contents in the garage to be cooled is:

$$W_{contents} = 500 \, lb$$

If we wish to drop the content temperature by $\Delta T = 20\,°F$ in $1/2$ hour and with $C_{contents} = 1.0 \, BTU/lb\,°F$ for the contents, then:

$$Q_{cool/h} = W_{contents} * C_{contents} * \Delta T / t_h = 20,000 \, BTU/h.$$

It would require about a 1.7-ton A/C unit to remove this heat, since 1 ton = 12,000 BTU/h.

This assumes an insulated garage and there is no heat loss due to heat coming in or leaving.

A 12,000-BTU/h or 1-ton window unit was available and installed. On hot days, it only dropped the temperature $10\,°F$ in the uninsulated garage, so a 20,000-BTU/h unit certainly would have been better.

10.8.1 Summary

Sizing an air conditioner would seem to show that a 1-ton or 12,000-BTU/h unit was not large enough to cool an uninsulated two-car garage. A unit two times bigger was needed.

10.9 At What Speed Does a Locomotive Become De-railed?

This question came up after a series of recent fatal railway accidents occurred on curves. Consider Figure 10.9.1 that represents a locomotive traveling around a curve of radius R. At what speed will it roll onto its side?

A locomotive with a velocity (V) around a flat curve, meaning not banked, of radius (R) produces a centripetal force in the (F_o) direction, like a swinging weight on a string. This is the inertial force trying to stretch the string or in this case keep

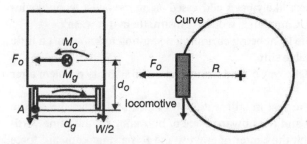

Figure 10.9.1 Train tipping.

the locomotive on the track. This is the same type force that moves you to the outer radius in a curve.

$$F_o = W/g^*a = (W/g)^*(V^2/R)$$

From the free-body diagram (FBD):

$$\Sigma M_A = 0$$

$$F_o^*d_o - (W/2)^*d_g = 0$$

Solving for the velocity to tip the locomotive:

$$V = (g^*R^*d_g/2^*d_o)^{1/2}$$

Assume $g = 32.2\,\text{ft/s}^2$, $R = 1{,}000\,\text{ft}$, $d_g = 4.7\,\text{ft}$, $d_o = 4\,\text{ft}$, then $V = 138\,\text{ft/s}$ (94 mph) and over that it will probably tip.

When a train is going 106 mph in that curve you know it will probably go on its side.

10.9.1 Summary

For most curves, a train tips at about 94 mph. High-speed trains follow larger curves so the tipping speed is much higher. Banking of the trackbed, which is not considered here, will increase this tipping speed.

10.10 Are Those Huge Cruise Ships Stable?

Whenever I take a cruise on a ship, this question always comes to mind. In a heavy sea, will the ship roll over on its side? These ships seem to be getting taller and taller as new decks are added to accommodate more passengers.

The concern comes from reading about capsized ferry boats that became unstable because of rough seas, shifting cargo, or ship weight modifications. Many such accidents occur every year.

These questions have most likely been addressed using the latest technologies on the newest cruise ships. Companies wouldn't insure them if it wasn't.

All of this doesn't stop me from being curious and wanting to know just a little bit more about what is a stable ship.

Figure 10.10.1 is a simplistic way of understanding a ship stability problem after doing a little research.

When a ship is floating at rest in still waters, it's acted on by two forces, the downward force of gravity and the upward force of buoyancy on the same vertical line. G is a force through the center of gravity, and B is a single upward force through the center of buoyance and is the result of the pressure on the hull.

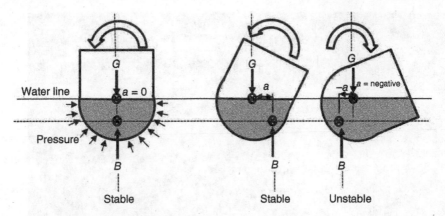

Figure 10.10.1 Ship stability from rolling over.

When a ship is disturbed either by wind or waves, it can incline and when this happens the volume of water under the hull shifts. This is shown in Figure 10.10.1. This causes the center of buoyance to shift also to the new geometric center of this mass. This causes a moment because of the moment arm (a) and this moment tends to restore the ship to its vertical position. This means it tries to restore the stability of ships. There are conditions that will make (a) negative that will make the ship unstable so it will try to "rollover." This is what ship architects design against.

The interesting fact is that the displaced volume is also equal to the dead weight of the ship. The heavier the ship, the further it sinks in the water and the stability may not change much, if the original design was sound. When more decks are added like in some ferries that capsized, little weight was added so the ship was unstable and rolled over.

While this doesn't say the big cruise ship I sail on next meets these criteria, I will now feel a little better sailing on it.

10.10.1 Summary

The height of a cruise ship isn't the only thing that defines its stability or its resistance for "rolling-over." The taller it is, the heavier it tends to be and the more water it displaces. This is in the direction of maintaining stability. In addition, recent ship designs incorporate sophisticated stability control devices.

10.11 Why Are Arches Used?

Sitting in church one Sunday, the beautiful laminated arch and triangle design supporting the roof as shown in Figure 10.11.1 were being admired.

Figure 10.11.1 Arches and triangles St. Martha's church.

Arches have been with us since antiquity and are used for strength. In the case of this new church, it allowed a high dome (i.e. 100 ft) with no center supports required in this large space. The Romans made maximum use of them in their aqueduct systems. Later, they made spectacular open churches possible.

All I wanted to do was explain the strength part to someone interested in more detail on how they worked and used Figure 10.11.2 along with some algebra.

From Figure 10.11.2, we see three transitions to the arch roof. The first is a flat roof, then the triangular roof and finally the circular or arched roof. They are all the same base width, but there is no height to the flat roof. The triangular roof and the arch roof are shown to have the same height. Consider the arch to be a half-circle i.e. ($h = L/2$) with an area $= (\pi/8)*L^2$ and the triangle has an area $= L^2/4$. Then for the case shown, for a given depth, the arch contains 57% more volume.

However, the primary benefit of the arch is that the load on the columns is in pure compressive stress. This means there are no dangerous bending tensile stresses to fail them, as there are in the flat roof.

The analysis of a variable arch can be fairly complex. For the purpose of an explanation, the arch will be considered to be similar to a triangle to show the compressive stress behavior.

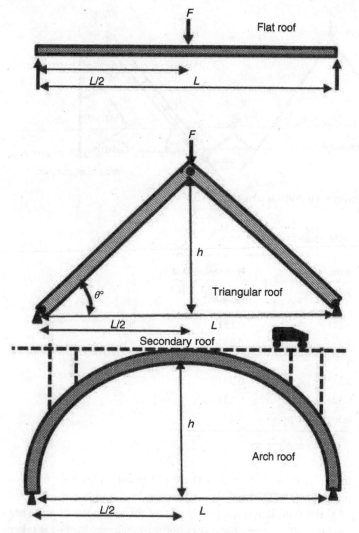

Figure 10.11.2 Flat, triangular, and circular roofs.

When the ends are fixed but free to rotate, the triangle reaction loads can be resolved into an axial force along the line of the triangle and a reaction force perpendicular to this axial force.

The relationship for the triangle is shown in Figure 10.11.3:

$$F_{axial} = F/2^*(\cos(90 - \theta) = F/2^*(\sin\theta)$$

Inserting some values for θ:

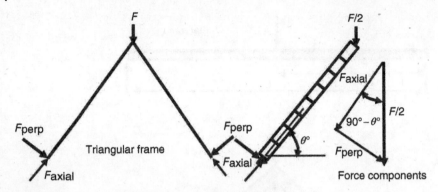

Figure 10.11.3 Forces on triangular structure.

Table 10.11.1 Triangle angle and reaction loads.

Angle	Axial load	Horizontal load
$\theta°$	$F/2 \cdot (\sin \theta)$	$F_{axial} \cdot \cos \theta$
15	0.13 F	0.13 F
30	0.25 F	0.22 F
45	0.35 F	0.25 F
60	0.58 F	0.30 F
75	0.52 F	0.13 F
90	0.50 F	0.0 F

It's interesting to see in Table 10.11.1 that there is a point where the shape of the triangle (θ) produces the highest axial compressive load and horizontal load in the member and at the foundation support point as shown on the chart. That is somewhere around $\theta = 60°$. Arches have this optimum shape also but it takes much more mathematics and testing to determine it.

Since the components are in compression, there will be little deflection or tensile stress so a longer span without mid supports is possible. Unfortunately, the horizontal forces are also maximized and must be properly supported at the base.

In Figure 10.11.2 of the arch, a secondary roof or road can be tied into the arch as shown. This allows the roof or roadway to be fairly flexible since the arch is supporting the load.

So the arch keeps the loads in compression and transfers the forces as thrust forces at the ground level generating high forces with higher loads. This is why the foundations of such structure have to be solid.

Churches such as Notre Dame in France had to add external braces called flying buttresses outside of the church. The purpose of the buttresses was to resist the horizontal forces pushing a wall outward by redirecting them into the ground.

10.11.1 Summary

Arches use compressive forces to resist failures in tension. They direct forces to the ground to be resisted there. This also eliminates the need for internal supports in a building. This increases the volume and the viewing area.

10.12 Why Don't Bighorn Sheep Die When Banging Their Heads?

Shown in Figure 10.12.1 is a sketch of bighorn sheep that have a ritual of banging their heads together at up to 30 mi/h (44 ft/s). They do this many times a year. The apparent G force on their brain, which is a little larger than ours, can be approximated using the analysis method of Section 8.1.

First consider no padding or $s = 0$ in.;

$$V = 44 \text{ ft/s}, s = 0 \text{ ft}, W_{\text{brain}} = 3 \text{ lb}, g = 32.2 \text{ ft/s}^2, \delta_{\text{cg}} = 0.042 \text{ ft}, d = 5.5 \text{ in}.$$

$$G's = 1,430$$

$$F_{\text{brain}} = 4,300 \text{ lb}_{\text{f}}$$

$$\sigma_{\text{brain}} = 181 \text{ lb/in.}^2$$

With such high forces and stresses, why don't these rams immediately die due to ruptures in the brain and broken skulls? Watching videos of this headbutting

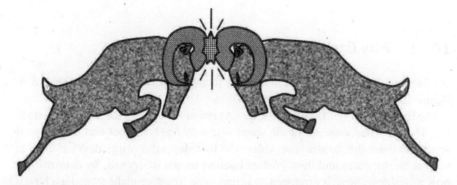

Figure 10.12.1 Bighorn sheep ramming.

shows that they seem to be a little dazed. One source says they rub their head on the dirt to relieve the pain. So it does hurt, but they do live for 8–10 years and they do the head butting throughout most of it. I don't know about the deterioration of their brain but they seem to be able to navigate the cliffs on the mountains without becoming disoriented and falling off. Females live a little longer and don't butt heads.

Reading up on these male rams reveals some interesting data:

- They have two skull bones that act as shock absorbers.
- Their large curled horns deflect and act as shock absorbers.
- Their skulls have a honeycomb type structure that absorbs shocks.
- The brain is more dense and fits tightly in the skull with little movement at impact.
- Thick tendons link skull and spine for support like a safety belt of sorts.

This shows cushioning is important as was shown in Section 8.1. Consider all of the above adds $s = 6$ in. of equivalent padding then:

$$G's = 110$$

$$\sigma_{brain} = 14 \, lb/in.^2$$

This is in the range of a mild concussion or being dazed.

10.12.1 Summary

Their horns and brains are designed to absorb the shock and act as padding as in a football helmet. No one knows if neurological damage to the brain occurs over time. The science of copying nature is the field known as biomimetics. Such an analysis on these rams suggests the possibility of a new bumper system.

10.13 Why Can't We Walk on Water?

While watching a video [1], it showed a lizard running on water as shown in Figure 10.13.1.

The lizard was about 1/2 lb, 6 in. long and ran at about 5 ft/s for 20 ft before sinking. The front feet were out of the water and only one back foot was in the water at a time. From the slow-motion video, the foot slapped down to develop force to keep it on the water and then pushed back to propel it forward. By determining how much downforce is produced to support the lizard's weight some idea how it stays on top of the water can be approximated.

Drawing by Allyson Sofronas

Figure 10.13.1 Lizard running on top of water [1].

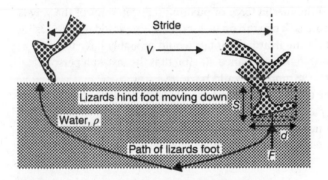

Figure 10.13.2 Lizard on water analytical model.

Figure 10.13.2 is the simple model. Notice the path the hindfoot takes as it pushes down to provide the lift force and then back to propel forward. All of the nomenclature used in the analytical model is shown in Figure 10.13.2.

The weight of water displaced downward (s) by the lizard's hindfoot:

$$W = (\pi^* d^2/4)^* s^* \rho \, \mathrm{lb_f}$$

When the forward velocity (V) in./s is known, the time t (s) for the downward motion s can be estimated as:

$$t \approx X \text{ stride (in.)}/V \text{ in./s}$$

The acceleration of this weight downward is due to the lizard's foot of diameter (d).

A slow-motion video [1] suggests that (s) is about equal to (d):

$$s \approx d$$

$$a = 2^*s/t^2 \text{ in./s}^2$$

$$F = (W/386)^*(a) \text{ lb}_f$$

which is the reaction force F (lb$_f$) on the water of the lizards foot.
Using the values,

$$V = 60 \text{ in./s (5 ft/s)}, d = 1.25 \text{ in., stride } (X) = 2.0 \text{ in.}, \rho = 0.036 \text{ lb/in.}^3$$

$$F = 0.32 \text{ lb}_f$$

which is only for the initial slap on the water. The rest of the support to carry the total 1/2 lb weight comes from the continued through motion.

10.13.1 Summary

The forward velocity and the inertia force of pushing the water out of the way is enough to support the lizard so it can run on water for a short period of time. Using the same model as used for the lizard, a human would probably have to travel at more than 90 ft/s (60 mph) to have a chance. Seeing that the fastest a person (e.g. U. Bolt) has gone is 41 ft/s (28 mph), it would be quite a feat.

Some researchers at the time of this writing are building small robots to perform this feat and have been partially successful. The power, weight, and stability are the primary factors that need to be optimized. The usefulness would be a device that can move over any terrain. Here is another case where engineering follows nature.

Reference

1 YouTube, Video by T. Hsieh, Lauder Laboratory, Harvard University, 2017.

10.14 How to Predict the Outcome of the Stock Market

This section was done to see how you could make money in the stock market. It's really like gambling and shouldn't be done with money you can't afford to lose.

During the writing of this book, the market took a real tumble because of the coronavirus pandemic. Most of our savings are in mutual funds and it's almost heart stopping to watch your lifesavings disappear. From past experiences, I'm somewhat consoled in knowing they are sound investments. By just doing nothing they will eventually rebound. This has happened a couple of times in the past. How long the downturn will be is not known but they always have returned and surpassed their original value. The worst thing to do is to panic and sell because then you have really lost your money and there's nothing to bounce back. The following was written before the coronavirus but the data has been updated.

Obviously, I've never learned how to pick a stock and sell it at the most opportune time or I'd be a very wealthy person. I've tried it and was never very successful.

I finally went into a diversified portfolio of stocks, bonds, gold, housing, cash, and overseas investments that have done well during our retirement. Our financial advisor manages the portfolio and keeps it balanced.

This doesn't mean I can't use a little research and mathematics to see if I can try to anticipate what the market will do on paper. During my career, I've used probability, statistics, vibration analysis, and sudden dynamic inputs to analyze problems. I don't want to get that technical here and want to keep things very simple by looking at historical data. Now the saying from investment guides is that history can't predict the future in the stock market and this I agree with. However, the following has been noticed:

- Sudden events that shake up the market can impact a stock price either up or down several days later.
- Some stocks may be impacted such as energy stocks and others may not such as technology stocks and it depends on the type event that has occurred.
- The duration or how long it takes them to reach their peak depends on the magnitude of the impact. A major event like a presidential election will have more effect than a statement someone makes on the economy.
- The stock market doesn't like uncertainties that could affect them and usually reacts to them by either raising or lowering the stock's value.
- No one can predict the future or the exact time a catastrophic occurrence will occur.

Figure 10.14.1 is an extremely simplistic model that will be used to collect data on a stock's price.

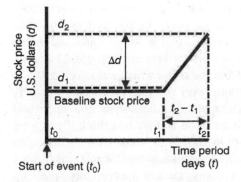

Figure 10.14.1   Stock price model.

All of the data on Figure 10.14.1 is available on the Internet on a company's stock price performance over time and the nomenclature is:

- d_1 is the current stock price in US $ at some time after the event (t_1).
- d_2 is the stock price in US $ at some time period after the event (t_2).
- t_0 is the zero time the event impact occurred or the first day.
- t_1 is the day after the event impact occurred.
- t_2 is the time the event impact measurement period ends.

Since we can't predict when a significant event will occur, we can't buy before the event occurs. All we can do is buy a stock after the event occurs, let's say the next day and see what its value will be in two months. To see if this works, we can look through historical stock values and see how they were impacted after a significant event occurred. Did the stock price go up or did it go down?

The real unknown is how the seriousness of an event will impact the stock price. About the only way to do this is to just obtain data on many stock situations and see what the impact did to them two months after the extreme event.

Notice that we are not trying to anticipate when an extreme event will occur. What we are doing here is saying the event has occurred and we have a day to buy the stock that is being tracked. Now once a stock drops in value, it's a bad idea to sell. Once you sell you have truly lost your money. By not selling you haven't lost anything, except on paper, since you still own the stock and it could rise in value. If it goes up and you sell, you may have lost if it were to rise further. However, if the commission fees aren't too high, you can have made some money at the price you sold the stock at but lost out in future growth.

Stocks in all sectors of the market will behave differently so the various sectors should be included.

For example, let's look at the extremes of the 2016 and 2008 presidential elections on stocks right after the results were announced and then two months later. The other very extreme event is the catastrophic Twin Tower event of 9/11/2001 when the stock market opened again a month later. The coronavirus pandemic raised havoc with the market with the Dow dropping 30% in value, all flights stopped, businesses closed, and 20 million workers were unemployed. The tables shown are samples of the data collected and the sector which was affected. While many of the stocks have had they problems over the years, stocks such as G.E. are kept on the chart just to describe the method. It does show how important balancing your portfolio is and how stocks come and go. Just think if you were wise enough to buy Apple at $1 a share in 2001. You would have increased its value by more than 280 times today. I didn't even know what Apple was in 2001 and didn't have any money to invest anyway. Luckily, I eventually wised up (see Tables 10.14.1–10.14.4).

Table 10.14.1 Market impact coronavirus 2/14/2020.

2/14/2020–4/14/2020 Coronavirus	d_1 start	d_2 end	%Δ	$t_2 - t_1$ Δt
Exxon/energy	$61	$43	−30%	60 d
G.E/industrial	$13	$7	−46%	60 d
Apple/technology	$325	$273	−16%	60 d
Chase/financial	$138	$98	−29%	60 d
UPS/transportation	$106	$98	−8%	60 d

Table 10.14.2 Market impact trump 11/18/2016.

11/8/2016 Trump	d_1 start	d_2 end	%Δ	$t_2 - t_1$ Δt
Exxon/energy	$86	$89	+4%	60 d
G.E/industrial	$28	$32	+14%	60 d
Apple/technology	$109	$118	+8%	60 d
Chase/financial	$68	$86	+26%	60 d
UPS/transportation	$107	$115	+7%	60 d

Table 10.14.3 Market impact Obama.

11/8/2009 Obama	d_1 start	d_2 end	%Δ	$t_2 - t_1$ Δt
Exxon/energy	$73	$66	−10%	60 d
G.E/industrial	$18	$17	−6%	60 d
Apple/technology	$27	$30	+11%	60 d
Chase/financial	$42	$44	+5%	60 d
UPS/transportation	$54	$62	+15%	60 d

Now these major impact events were highly disruptive to the market and they certainly impacted various stock sector prices. Positive events like presidential elections caused most stocks to rise since the population was usually split 50–50% on who they wanted to become president. The problem is it affected different sector in completely different ways depending on what the investors thought would affect their segment of the market.

Table 10.14.4 Market impact Twin Towers 9/11/2001.

9/11/2001 Twin Towers	d_1 start	d_2 end	%Δ	$t_2 - t_1$ Δt
Exxon/energy	$38	$38	0%	60 d
G.E/industrial	$31	$41	+32%	60 d
Apple/technology	$1.1	$1.2	+9%	60 d
Chase/financial	$41	$39	−5%	60 d
UPS/transportation	$50	$54	+8%	60 d

The coronavirus was different since it totally shut down the world economy. It was worldwide and greatly affected trade and may be the worst change in the economic condition the United States has ever seen. The oil market crashed because it was suspected certain countries were trying to dry up the US shale oil business. Many small drilling companies in the United States are highly leveraged, meaning they have big loans. When oil prices are too low, because the market is flooded with low-cost oil they go bankrupt. Likewise, the high unemployment skyrocketed because of the government's order to self-isolate and social distancing. It was a rough time for the world. The market eventually came back after four months even though the coronavirus was still present.

I would imagine if you were willing to do this type of analysis for a wide variety of stocks you might be able to determine a trend. However, you must know what a major disruptive event is and know if it will drive the market up or down. That involves human emotions in the market or who controls the market such as computer selling. Only with a large amount of data collecting and research could you start to understand this.

For these reasons, I'll stick with a diversified portfolio with a financial advisor doing the balancing of the portfolio. I don't like all of my eggs in one basket and I can think of other more enjoyable ways of making money than sitting in front of a computer screen all day collecting data and worrying. Remember what we are doing here is considered speculation and not investing. You could lose it all.

10.14.1 Summary

The stock market seems to be too unpredictable to determine what will happen due to a sudden impact of an unforeseen event the day after it occurs, with the limited data collected. Mathematical analyses are good at assessing the risk but are usually a failure at predicting the future. With computerized algorithms, performing the buying and selling what individuals think seems less important.

10.15   Things Aren't as Random as They May Appear

During my career in industry, there has been a reappearing theme to the work that's been done, failures analyzed, or things noticed. The following tries to explain why this might be so.

I've often sat in a classroom of people and wondered how many have the same birthday as me. The same day, month, and year would be nice to know, but just knowing if someone in the group had the same day and month would be interesting too. Now over the years, I've used statistics and probabilities quite a bit but usually use existing equations or methods. Recently, I saw the simple derivation on the probability of two people in a group of 30 people, possibly a classroom, having the same birthday, meaning day and month [1].

So it first will be shown what the probability is that no one shares a birthday with anyone else $p(n)$. The probability that someone shares a birthday with at least one other person will then be determined and is $p(s)$. Consider Figure 10.15.1.

First let's consider three people in the group to show the trend:

The first person will have a chance of their birthday being 365 out of 365 days of the year or 1/365/365.

The second person can be born on any day the first person wasn't born on or 1/364/365 and the third person wasn't born on either of the other two person's birthdays, 1/363/365.

By the multiplication rule of probabilities;

$$1/365/365*1/364/365*1/363/365$$

This comes up with the probability no one sharing the same birthday as:

$$p(n) = 365*364*363/(365)^3$$

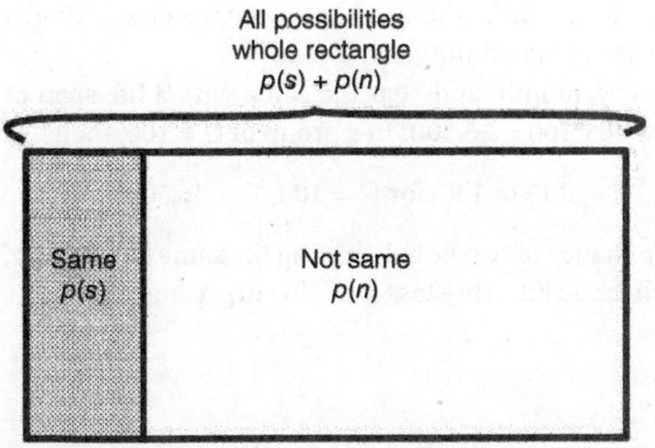

Figure 10.15.1   Figuring the probability.

Seeing this progression, the same method can be used for a group of 30 people instead of 3:

$$p(n) = 365*364*363* \ldots 336/(365)^{30}$$

$$p(n) = 365*364*363* \ldots 336/(365)^{30}$$

Now $365*364*363* \ldots 336 = 365!/335!$ or $365!/(365-30)!$

The probability of no one having the same birthday is therefore:

$$p(n) = 365!/(365-30)!/(365)^{30} = 0.294 \text{ or } 29.4\%$$

The probability of someone in the group of 30 having the same day and month birthday;

$$p(s) = 1 - p(n) = 0.70 \text{ or } 70\% \text{ which seems like pretty good odds.}$$

This can be developed as a generalized equation:

$T = $ total population

$G = $ group size

$$p(s) = 1 - (T)!/[T^G*(T-G)!]$$

This results in pretty large factorials that some calculators can't handle. A simple form using the Taylor Series and some mathematical gymnastics and using 2.71826 as the base of (e) is:

$$p(s) = 1 - 2.718^{[-G*G/(2*T)]} = 0.708 \text{ or } 70\% \text{ for } G = 30, T = 365$$

This would show that when the group size (G) is increased to 100, the probability of someone in the group having the same birthday $p(s)$ as someone else is almost certain (99.9%) since there are 365 (T) possibilities in a year.

Consider we want the same day, month, and year the same with a life span of 100 years. The pool or T is now $365*100 = 36,500$. In a group of $G = 100$, then:

$$p(s) = 1 - 2.718^{[-G*G/(2*T)]} = 0.13 \text{ or } 13\% \text{ for } G = 100, T = 36,500$$

So in a group of 100 people, the chance of two people having the same day, month, and year birthday is 13% which is considerably less but still surprisingly high.

Reference

1 Khan, S. (2009). *Birthday Probability Problem*. Khan Academy, YouTube.

10.15.1 Summary

In a group of 100 people, there's an extremely high chance that at least two people in the group will have the same day and month birthday. For a group of 100 people, the chance for the two people having the same day, month, and year birthday is about 13%.

10.16 Why Do Certain Events Seem to Happen Quite Often?

Something has happened to me over the last few years. For some reason, a few times per month, whenever I look at our digital clocks, they say 7 : 47. My wife has gotten tired of me saying, "Look at the time, it's 7 : 47." I was starting to think it was an omen and something was going to happen to me someday at that time. I then reviewed what was done in Section 10.15 and realized it wasn't an omen but a probability.

Think about it. A day has 24 hours but I'm only awake to look at the clock for 16 hours. The time 7 : 47 comes up twice during that time of 960 minutes. Using the methods of Section 10.15 with a group size of 2 and the possibilities of 960 says, there's less than a 0.2% chance of me seeing the number each day. Actually, I look at the clock about 10 times per day so it's more like a 2% chance of seeing the number.

This doesn't tell the whole story. I tend to center my observations at breakfast and when certain prime-time television shows are on. So the 960 minutes becomes more like 240 minutes and the 2% becomes 10%. This isn't a high probability but over a month seeing the 7 : 47 appear several times does seem reasonable.

10.16.1 Summary

A reoccurring number such as 7 : 47 on a clock is because there are only so many possibilities each day for that number to occur. For my case, it was a 10% probability. Over a month, its occurrence was a surprise because I was looking for it to happen. I may have even favored a time I would look at the clock.

10.17 Occurrences on Machines and Structures

As engineers we sometimes think we are working on the same type of problems over and over. There is a reason for this and that's because we probably are. That's why I'm always looking for unique problems to solve. I get bored doing the same thing several times.

One useful tool to describe this repeating scenario is called the Pareto Principle, which is also called the 80 : 20 rule. It's not a mathematically derived rule but more one developed by observations. It states that for many events, roughly 80% of the effects come from 20% of the causes. For example, Vilfredo Pareto (1848–1923) used it to show that 80% of the wealth of England was held by 20% of the population. I've noticed it in donations where about 70% of the contributions come from 30% of the members in a church group. It's used in many fields such as sports, quality control, and computing. Where I've noticed, it doesn't seem to correlate well is that being successful in business does not correlate with level of education. Creativity does not correlate with education. Some of the most skilled equipment designers I've known never completed high school.

In engineering all that will be shown here are my experiences in the areas of research, troubleshooting, and design. It is a useful relationship to make sure you don't miss reviewing causes with a high potential for failure. Many of the problems analyzed in this book fall into the 20% range since they were difficult to solve and the cause was not obvious. A detailed analysis may have been required to identify the true cause.

Consider centrifugal pump failures that have the following major failure causes:

1. Cavitation related problems
2. Lack of lubrication
3. Poor repairs
4. Wrong process lineup
5. Alignment
6. Defective design
7. Coupling failures
8. Bad bearings/inadequate lubrication
9. Seal failure
10. Undefined

When reviewing the failure history of 200 pumps at one plant location, items 2, 8, and 9 were experienced most often. It wasn't 80 : 20 but closer to 70 : 30. This simply means that these items should be considered in a failure investigation and corrections made so the failure won't be repeated.

Gear failures were a considerable part of failure investigations and consisted of approximately 50 failures of some type. They were usually due to one of the following causes:

1. Overloading from system torsionals or cold start-ups
2. Improper lubrication
3. Gear or gearbox design
4. Misalignment
5. Improper heat treatment
6. Poor quality control
7. Adverse operating conditions

In this case, most of the failures were due to 1, 2. This ratio was 50% of failures due to 33% of the causes. Not the 80 : 20 rule but still proves a point and that is to not overlook them. This sampling is highly dependent on the industry one works with or the area one concentrates their effort. I performed many torsional analyses, so the data is bias in this area.

Both ball and roller bearings were another area of concern and were usually failure areas.

1. Lack or contaminated lubrication
2. Improper assembly
3. Excessive loading
4. Defective bearing
5. Other

In this case, 75% of the failures analyzed were due to 60% of the causes primarily items 1 and 2 with 3 a close third. These should be considered in any failure analysis.

Centrifugal compressors also experienced failures; however, my experience base was only 20 failures and they were the nontraditional type that required an analysis. The small sampling was for my use only but is presented here because it covered several industries.

1. Damaged blades
2. Lubrication problems
3. Surge problems, controls
4. Alignment
5. Imbalance
6. Torsional vibration related
7. Shaft or coupling failure
8. Bearing failures
9. Seal failures

The number of failures analyzed was very limited with only 20 investigations. In this small sample, 70% of the failures were due to 30% of the items. Items 4, 8, and 9 were the major failures examined.

Reciprocating gas engine compressors also were investigated. The failures witnessed and addressed were in the following areas:

1. Valve damage
2. Bearing failure
3. Rider band failure
4. Rod seals
5. Alignment problems
6. Lubrication problems
7. Turbocharger failures
8. Miscellaneous

Figure 10.17.1 Doghouse slipper loosening.

Item 8, miscellaneous was usually the major contributor with items such as grouting deterioration, detonation due to fouled turbocharger coolers pounding out main and rod bearings. Item 1 was the second most experienced by this author. This is probably because the other failures were quickly repaired but specialists were called in for the high-profile failures. High profile is meant safety, liability, or production concerns for top management. This was about an 80 : 20 ratio and was where most concern was based on 40 compressors. Some of the uncommon type failures in large gas engine reciprocating compressors are like those of Figure 10.17.1 which occurred from loose bolts. Not a regular occurrence but they do occur in certain designs and greatly limit production.

10.17.1 Summary

Failures seem to happen in a regular pattern. There are just so many ways that events happen that you are bound to see some reoccurrence if you work in a certain area long enough. This can be used to your advantage as it will remind you not to miss analyzing certain possibilities.

The examples here are my experiences and usually are because I was called in to help define the cause so in that way it is not general for all machines or industries.

10.18 How Long Does It Take to Thaw a Frozen Turkey and to Cook It?

Every thanksgiving my wife asks me how long the 10-lb turkey will take to thaw out from the freezer and I wonder what it is for different weights and meats. Later comes the question of long should it be cooked.

Here is a place to use the specific heat (C Btu/lb °F) equation and the convection equation.

The heat rate lost due to the temperature difference in a given period of time (t_h) is:

$$q_{turkey} = W_{turkey} * C * (T_{freezer} - T_{refrigerator})/t_h \text{ Btu/h}$$
$$= 10*0.7*(0°F - 35°F)/t_h = 245/t_h \text{ Btu/h}$$

The convection equation is the rate of heat loss from the turkey (W_{turkey}) to the environment due to free convection.

$$q_{convection} = h * A * (T_{freezer} - T_{refrigerator})$$

The heat transfer coefficient for still air and is about $h = 0.2$ Btu/h ft² °F and A is the turkey surface area exposed to the environment in the refrigerator or about 1.0 ft^2

$$245/t_h = 0.2*1.0*(-35)$$

Equating and solving for t_h:

$$t_h = 35 \text{ hours to get from 0 to } 35°F \text{ or about 3.5 h/lb.}$$

A follow-up question to this is how long the 10 lb turkey will take to heat up when put in the oven from the refrigerator.

The same specific heat equation and the convection equation are used for this.

The heat rate input due to the temperature difference in a given period of time (t_h) is:

$$q_{turkey} = W_{turkey} * C * (T_{refrigerator} - T_{165})/t_h \text{ Btu/h}$$
$$= 10*0.7*(35° - 165°)/t_h = -910/t_h \text{ Btu/h}$$

The convection equation is the rate of heat gain to the turkey (W_{turkey}) from the oven due to free convection.

$$= h * A * (T_{refrigerator} - T_{oven}) \text{ Btu/h}$$

where h is the heat transfer coefficient with some heating turbulence and is about 1.0 and A is the turkey surface area exposed to the environment in the refrigerator or about 1.0 ft^2

$$q_{convection} = 1.0*1.0*(-315) = -315 \text{ Btu/h}$$

Since $q_{turkey} = q_{convection}$ then $-910/t_h = -315$

$$t_h \approx 3.0 \text{ hours}$$

10.18.1 Summary

It takes about 3.5 hours/lb of frozen meat to thaw out when taken from the freezer and put in the refrigerator section.

It takes about 1/4 hours/lb from the refrigerator at a 32 to 315 °F oven to get the meat to 165 °F.

11

Magic Tricks Using Engineering Principles

Engineers Are Difficult To Mystify As Their Livelihood Is Solving Complex Problems

11.1   Surface Tension and Floating Metal

While the lizard uses its momentum to run on top of water as shown in Section 10.13, insects such as the water strider shown in Figure 11.1.1 use surface tension.

Surface tension is related to molecular forces and causes a surface film that makes it difficult to move an object through the surface. It's because the water molecules on the surface move more freely than the interior molecules and can hold together, which is called cohesion. The surface molecules have no water molecules above them. Interior molecules are restricted by the other water molecules around them. It's as if the surface was covered by a thin stretched membrane.

A steel paper clip ($\rho_{steel} = 0.28\,\text{lb/in.}^3$) is denser than water ($\rho_w = 0.036\,\text{lb/in.}^3$) but it can float on top of the water as in Figure 11.1.2.

My experience doing this is that your fingers need to be very steady and the dry paper clip to maintain the film. Archimedes' principle says this shouldn't be so. It also says that a body immersed in a fluid produces an upward force equal to the weight of the fluid displaced. Since the paper clip weighs more than the fluid it displaces it should sink, but it doesn't. This is because of the surface tension and because the paper clip is never immersed. It lays on top of the film. From the experiment, it shows it can float.

The weight this surface tension can support (W_{st}) on the surface of still water is determined from experiments where L_{in} is the length in contact with the water:

$$W_{st} \approx 0.0004 * L_{in}\ \ \text{lb}_f$$

Unique Methods for Analyzing Failures and Catastrophic Events: A Practical Guide for Engineers, First Edition. Anthony Sofronas.
© 2022 John Wiley & Sons, Inc. Published 2022 by John Wiley & Sons, Inc.

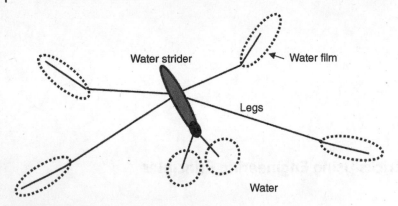

Figure 11.1.1   Water strider on a water film.

Figure 11.1.2   Floating on water due to surface tension.

The paper clip weighs $W_{steel} = 0.0011\,\text{lb}_f$ and unrolled length is $L_{in} = 3.25\,\text{in.}$ long.

Surface tension can support:

$$W_{st} \approx 0.0004 \times 3.25 = 0.0013\,\text{lb}_f \text{ so it floats on the film.}$$

An article shows a coin floating on the water, so a copper penny was tried and it wouldn't float. However, a Japanese Yen does float. It's 100% aluminum and has a density of $\rho_{\text{alum}} = 0.1$ lb/in.^{3} which is about three times heavier than water so it should also sink but it doesn't.

I wanted to use the same equation but didn't know what L_{in} should be as it represents a line contact.

Maybe the area would be the thing to use to balance the 0.0022 lb (i.e. 1 g) Yen coin weight and 0.787 in. (i.e. 20 mm) diameter coin.

The surface tension can support using line contact:

$$W_{\text{st}} \approx 0.0004^*(\pi/4)^*(0.787)^2 = 0.000195 \text{ lb}_{\text{f}}$$

Clearly, the constant 0.0004 must be greater to just balance the coin when a surface area is used.

So modifying the constant of the equation when a surface area is used is:

$$W_{\text{st}} \approx 0.0045^*(\pi/4)^*d^2 \approx 0.0045^*(\pi/4)^*(0.787)^2 \approx 0.0022 \text{ lb}$$

so it just balances the weight and keeps from falling through the surface film as shown in Figure 11.1.3. Notice the lip showing it is floating on the film and not into the fluid.

What about the US Lincoln penny, the weight of which is 0.0055 lb (i.e. 2.5 g) and 0.75 in. (i.e. 19 mm) diameter? It doesn't float no matter how hard one tries.

$$W_{\text{st}} \approx 0.0045^*(\pi/4)^*(0.75)^2 \approx 0.0036 \text{ lb}$$

so the surface tension of plain water will never be enough to support the coin on the water. If a magician floats a penny, it is made of some other material than a standard penny, has been lightened or something has been added to the water. Physics shows it isn't possible to float it without altering it.

Figure 11.1.3   Yen floating on water.

11.1.1 Summary

Surface tension is like a thin membrane and can support objects on top of the film if certain conditions are met. This is true even if it seems to violate Archimedes' principle.

11.2 Acceleration of Gravity and the Money Challenge

This is based on how quickly something accelerates under the pull of gravity.

Figure 11.2.1 illustrates how you should hold a dollar with your fingertips. The person you are challenging puts their fingers as is shown, about 1 in. from the top, not touching the bill. The challenge is for them to pinch their fingers together and catch the bill after you release it before it passes through. Most can't catch it and they will be frustrated and amazed after several tries.

Figure 11.2.1   Initial position for catching the bill.

The distance (*s* [in.]) an object falls under the acceleration of gravity ($g = 386\,\text{in/s}^2$) in a given time period (*t* [s]) with no air friction is represented by the well-known equation:

$$s = 1/2\,g^*t^2$$

The key to this trick is that the average human reaction time is $t = 0.2$ seconds. This means it takes that long for them to squeeze their fingers together. In this amount of time, the dollar bill has moved down:

$$s = 1/2\,g^*t^2 = (1/2)^*(386)^*(0.2)^2 = 7.72\,\text{in.}$$

This means the bill is well beyond the grip point.

This simple equation will also allow you to determine how deep a water well is with or without water. Just measure the time (*t*) after you release a stone and listen for the splash or thump. Use the equation to determine the depth (*s* [in.]).

11.2.1 Summary

The fingers can't react fast enough to catch the dollar bill as it slips through their fingers and it drops to the ground.

11.3 The Jumping Coin

This challenge will ask someone if they can get a coin to jump into a cup without using their hands. The answer is yes, if you blow on it precisely in the right amount. Figure 11.3.1 illustrates this being done.

The physics to explain this is based on Bernoulli's principle as shown in Section 12.11.

This says that the higher the velocity, the lower the pressure. When the pressure times the coin's surface area over the coin is greater than the coin's weight, it will lift up and be blown into the cup.

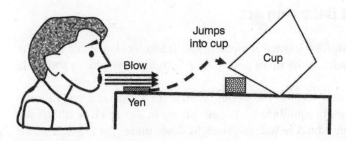

Figure 11.3.1 Jumping a coin no hands.

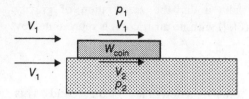

Figure 11.3.2 The Bernoulli lift model.

Simplifying Bernoulli's equation and using the model of Figure 11.3.2,

$$p_1/\rho + V_1^2/2g = p_2/\rho + V_2^2/2g \text{ and since } p_2 \text{ and } V_2 \text{ are zero;}$$

$$p_1 = -\rho^* V_1^2/2g$$

The force trying to lift the coin is $F_{coin} = A_{coin}{}^* p_1$
The coin will lift or jump when $F_{coin} > W_{coin}$
When they are just equal $p_1 = W_{coin}/A_{coin}$
Solving for the velocity, one will need to blow to make them equal:

$$V_1 = [2^* g^* (W_{coin}/A_{coin})/\rho]^{1/2} \text{ in./s}$$

$$W_{coin} = 0.0022 \text{ lb}, A_{coin} = 0.486 \text{ in.}^2, \rho = 4.5^* 10^{-5} \text{ lb/in.}^3, g = 386 \text{ in./s}^2$$

$$V_1 = 277 \text{ in./s or } 18.9 \text{ ft/s}$$

With a digital anemometer, a blast of air from the mouth at 40 ft/s or more is possible and is certainly enough to jump and push a 1-Yen coin that was used here. Heavier coins take higher velocities.

11.3.1 Summary

The coin will jump because there will be less pressure on top and the coin will lift off. The velocity has to be high enough to do this. Most trying to do this won't give a blow burst hard enough to get the coin into the air and then into the cup.

11.4 The Belt Balancing Act

This was a little magic trick learned at a young age. It involves balancing a weight, in this case a stiff belt, on the tip of your finger with a clothespin, as shown in Figure 11.4.1.

On the Internet, many sites say it works because the center of gravity rotates under the finger so it's in equilibrium. As one can see in Figure 11.4.1, this is not totally true or the belt would be hanging straight down under the finger. There is

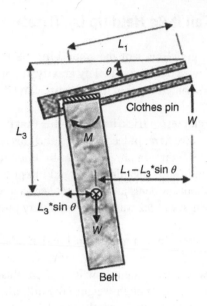

Figure 11.4.1 FBD of clothespin trick.

more to it because there are also restoring moments on the clothespin. These are trying to hold the clothespin straight, so it rotates to a certain angle (θ) until the moments and forces balance. They must be in equilibrium so they won't move. The free-body diagram (FBD) has been simplified so the effect of the moment (M) is shown.

Summing the moments and forces results in:

$$\sin \theta = L_1/(2^*L_3)$$

For the case shown in Figure 11.4.1

$$L_1 = 4 \text{ in.}, L_3 = 10 \text{ in.}$$

So the belt will be in equilibrium when $\theta \approx 12°$.

11.4.1 Summary

The belt doesn't fall and stays somewhat horizontal because of the reaction of the equilibrium moments within the clothespin. The clothespin finds a position such that the forces are in equilibrium.

11.5 How Can It Be Held Up by Threads?

My nine-year-old grandson sent me something he had built and is shown in Figure 11.5.1, a device that seems to defy gravity. It appears that there is nothing holding it up as it's something thin threads sure can't do it. Have you ever tried to push on a thread?

After doing the clothespin trick, it was obvious that there has to be some equilibrium equations involve here. Indeed, there is and this is the field of pretensioning called Tensegrity Structures, a phrase coined by Buckminster Fuller in the 1960s. It's an abbreviation of sorts meaning tensional integrity. This subject is discussed in engineering schools. Models are fun to make, so after a lot of aggravation the model shown in Figure 11.5.2 was built. Not a very pretty design but it did prove the concept.

I was about to develop a three-dimensional model to explain all the forces and moments internal to the structure. However, after building the model of Figure 11.5.2, it was quite evident how it worked. Push down on the top of the structure, meaning apply enough weight (W) and all the outer threads labeled (A) went limp, but it would still carry the weight.

The key to understanding this is seen in Figure 11.5.2. The upper arm as shown acts as a cantilever spring when it is bent down and when released pulls up on the bottom arm. This puts all the strings in tension.

When a weight (W) is put on top that is large enough, it will still pull on the center thread (B) but the A threads will become slack. Even without the preload in thread (B), it will still carry the tension load caused by the weight.

Built by Anthony N. Sofronas

Figure 11.5.1 Antigravity device.

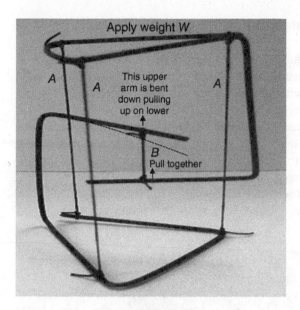

Figure 11.5.2 Verification model.

When the unit was built, the center thread (B) was pulled tight between the two arms and when tied and released it pretensioned the whole structure and all the other threads became tight.

With the model built, several things were evident:

- All the threads were in tension since a thread can't have compressive forces.
- The tighter thread (B) was, the increasingly rigid structure became.
- The only thing that made the structure collapse was if it was twisted out of alignment or one of the strings broke.

11.5.1 Summary

It was apparent that the threads in the structure were all in tension caused by the preload of the center thread. This preload and the geometry are what made the structure rigid enough to carry weight. This small model had a preload of about $(b) = 1/3\,\text{lb}_f$ and could carry a weight of $(W) = 1/3\,\text{lb}_f$ and each thread carried $(a) = 1/9\,\text{lb}_f$. This is what made the structure look like it was levitating.

This type of structure had no redundancy since loss of the preload in the (B) center thread would cause the whole structure to collapse. Likewise, loss of one of the (A) threads would cause the structure to twist and collapse.

The model of Figure 11.5.1 can carry more weight than the scissor.

11.6 Pulling the Tablecloth

We've all seen this being performed, sometimes with disastrous results. A table is completely set with dishware and silverware on a tablecloth. Your favorite uncle grabs the end of the tablecloth, pulls it off the table quickly, and the dishes and silverware stay in place. Everyone is amazed except of course your mom who is just angry at her brother. When no one is home you try it and thus the disastrous results. So what happened? Figure 11.6.1 illustrates the trick and the arms quickly pulling down on the tablecloth so it won't lift up.

Figure 11.6.2 is the model used for the analysis.

The frictional force under the dish is $F_{friction} = \mu * W_{dish}$ and if you pulled on the tablecloth, the dish would like to come along due to fiction between the plate and cloth.

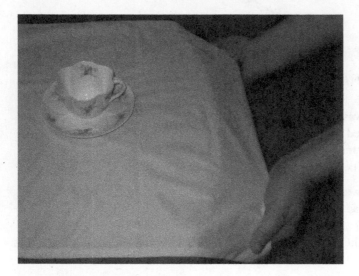

Figure 11.6.1 Tablecloth trick.

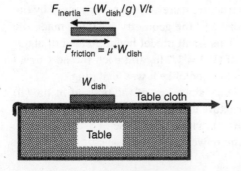

Figure 11.6.2 Tablecloth model.

Accelerating this pull and the dish inertia wants to resist moving. The force required to accelerate W_{dish} is:

$$F_{inertia} = W_{dish}/g^*a = (W_{dish}/g) V/t_s$$

V is the speed of the tablecloth if it is pulled in (t_s) seconds.

At some point, the resisting force to accelerate will be greater than the force to pull the dish with friction:

So to not move $F_{inertia} > F_{friction}$ or the inertia force is greater than the friction force.

$$(W_{dish}/g) V/t_s > \mu^* W_{dish}$$

When this inequality holds, the dish should remain on the table.

Assume $\mu = 0.5, V = 3\,\text{ft}/t_s = 10\,\text{ft/s}, t_s = 0.3\,\text{s}, g = 32.2\,\text{ft/s}^2$

1.04 > 0.5 and since the inequality holds the dishes should stay on the table.

11.6.1 Summary

This shows that it isn't just Newton's Laws as described in Chapter 12, based on inertia that is responsible for the successful outcome of this trick. It also involves friction. When the tablecloth is pulled flat with no hems on the edges in about 0.3 seconds or about as fast as you can pull 3 ft of tablecloth, the dishes will stay on the table.

Hesitate when you pull, pull unevenly, pull too slowly, or have seams on the tablecloth or dishes that can grip the tablecloth and broken dishes will result.

12

Useful Forms of the Equations Used in this Book

The Foggy First Look At A Failure Clears Up With Good Data

12.1 The Equations of Motion

For constant acceleration or deceleration (change signs for deceleration):

$$t = (V_f - V_i)/a$$

$$d = V_i{}^* t + 1/2\, a^* t^2$$

$$a = V_f^2 - V_i^2 / 2^* d$$

$$V_f = V_i + 2^* a^* t$$

$$d = ((V_f + V_i)/2)^* t$$

(See Table 12.1.1).

12.2 Newton's First Law of Force

An object at rest will remain at rest unless acted upon by an unbalanced force. An object in motion continues in motion with the same speed and in the same direction unless acted upon by an unbalanced force. This law is often called "the law of inertia."

This means unless something like friction, gravity, drag, or an opposing force is trying to stop something it will keep on going.

Unique Methods for Analyzing Failures and Catastrophic Events: A Practical Guide for Engineers,
First Edition. Anthony Sofronas.
© 2022 John Wiley & Sons, Inc. Published 2022 by John Wiley & Sons, Inc.

Table 12.1.1 Nomenclature and typical units.

V_i	Initial velocity	ft/s
V_f	Final velocity	ft/s
D	Distance traveled	ft
A	Acceleration or deceleration	ft/s^2
T	Time	s

12.3 Newton's Second Law of Force

Acceleration is produced when a force acts on a mass. The greater the mass of the object being accelerated, the greater the amount of force needed to accelerate the object. The equation is $F = m*a$.

This means that if you try to accelerate or decelerate something, then this is the force it takes.

This can be written using G's where $G's = a/g$

$$F = W^* G's$$

12.4 Newton's Third Law of Force

For every action, there is an equal and opposite reaction.

This means if you push on a rigid wall, it pushes back just as hard.

12.5 Newton's Gravitation Theory

This theory states that every mass attracts every other mass by a force pulling the two masses together. This force is inversely proportional to the square of the distance between them.

In equation form for gravitational force:

$$F = G^* m_1{}^* m_2/r^2$$

where F is the force between the mass, m_1 and m_2 are the masses, and r is the distance between them. G is a gravitational constant. This is a constant value and remains this value throughout the universe. This "big G" should not be confused with the local gravitation constant on Earth "little g" $= 32.2\,\text{ft/s}^2$.

The gravitational field intensity, also known as gravitational acceleration, is the acceleration of any mass due to the gravitational field and in equation form is G^*M/r^2. The gravitational field requires only one mass, while two masses are required for gravitational force.

12.6   Static Equilibrium

When all the forces that act upon an object are balanced, then the object is said to be in a state of static equilibrium.

$\Sigma F = 0$ means the sum of all forces on the structure is equal to zero.

$\Sigma M = 0$ means the sum of all moments about some point is equal to zero.

When there are only two unknowns and when these two equations are used, the problem is said to be statically determinate. With three are unknowns, the problem is defined as statically indeterminate and another equation would need to be used, usually deflection.

12.7   Momentum and Impulse

Momentum = mass times velocity.

Impulse = force times time acted on.

Conservation of momentum is a useful equation:

$$m_1 V_1 = m_2 V_2$$

Another useful form of these two equations comes from Newton's Second Law

$$F = m^* a$$

But $a =$ velocity (V)/time (t).

So inserting and rearranging:

$$F^* t = m^* V \text{ or } F = m^* V / t$$

Knowing the duration (t) of the force and the velocity (V) of the mass (m), the force (F) on the mass can be determined.

12.8   Kinetic Energy

This is the energy contained in a moving body. An automobile of mass (m) traveling at velocity (V) contains kinetic energy (KE).

The equation is:

$$KE = 1/2 \, m^* V^2$$

12.9   Potential Energy

Potential energy is just that, the potential for doing work, while kinetic energy is work being done. If you drop a 100-lb rock (W), while it is falling it possesses kinetic energy but as you are holding it at a certain height (h), it contains potential energy.

The equation for potential energy or potential work due to height is:

$$PE = W^*h$$

A spring also contains potential energy and when k lb/in. is the spring constant and δ is the compression of spring, the PE is:

$$PE = 1/2^*k^*\delta^2$$

12.10   Conservation of Energy

Energy can neither be created nor destroyed, rather it transforms from one form to another.

This is evident when a car goes to the top of a hill. Driving up to the top, it is kinetic energy doing the work but up at the top of the hill and stopped it now possesses potential energy. Likewise for heat is conducted out of a body and is then converted into convected heat. Water flow into a pipe is equal to water flowing out of a pipe as long as it isn't stored. Power in is equal to power dissipated is another form.

12.11   Bernoulli's Equation

In fluid dynamics, Bernoulli's principle states that an increase in the speed of a fluid occurs simultaneously with a decrease in pressure or a decrease in the fluid's potential energy. The principle is named after Daniel Bernoulli.

The Bernoulli equation is very useful, partly because it is very simple to use and also because it can give great insight into the balance between pressure, velocity, and elevation. For various situations, many of the terms in the equations drop out. For example,

$$Z_1 + V_1^2/2g + p_1/\rho_1 = Z_2 + V_2^2/2g + p_2/\rho_2 + \text{losses}$$ when the flow is at the same elevation then $Z_1 = Z_2 = 0$ and with no losses, losses = 0 (Table 12.11.1).

Table 12.11.1 Nomenclature and typical units.

V_1, V_2	Velocities	ft/s
p_1, p_2	Pressures	lb/ft^2
Z_1, Z_2	Elevations	ft
G	Gravity constant	ft/s^2

12.12 Specific Heat Equation

This is the amount of heat Q required to raise a weight of material (W) 1 °F, where C is the specific heat of the substance. For example, C for water is 1:

   BTU/(°F lb), for air 0.24, and for steel 0.12

$$Q = W_{lb}{}^*C^*(T_2 - T_1) \text{ Btu}$$

A convenient form for flows of liquids or gases is where w is the weight flow (lb/h):

$$q = w^*C^*(T_2 - T_1) \text{ Btu/h}$$

12.13 Conduction Equation

For one-dimensional heat transfer, say along the x-axis only, Fourier's law of heat or conduction through the material is:

$$q = k^*A^*(T_2 - T_1)/l \text{ Btu/h}$$

where q is the heat flow through the material of thickness (l) and the area (A), with the temperature differential ($T_2 - T_1$) and (k) is the thermal conductivity of the material available from test data.

12.14 Convection Equation

For one-dimensional heat transfer, say along the x-axis only, Newton's law of cooling or convection is:

$$q = h^*A^*[T_2 - T_1] \text{ Btu/h}$$

Notice the similarity to the conduction equation only (k) is now (h) which is the boundary layer heat-transfer coefficient. It can be for free convection as in still air or forced convection like a fan would produce on the surface.

12.15 Radiation Equation

The radiation equations is complex but can be simplified for a small body radiating to space, like a steam pipe in a large room or a person to the environment.

$$q = \varepsilon^* \sigma^* \left[(460 + T_2)^4 - (460 + T_1)^4 \right] \text{ Btu/h}$$

The use of the equation along with constants (ε) and (σ) is provided in the examples.

12.16 Theories of Material Failure

This is defined here as how the material failed. The type failures are fatigue where you cyclic stress something until it fails, like bending a metal paper clip back and forth. It eventually causes a crack or failure. A shear failure is when it fails in shear-like punching a hole in metal. A yield failure is when a metal part is bent and doesn't spring back to its original position. Another is a tensile failure when the part doesn't spring back but breaks, like bending a piece of blackboard chalk that is also a brittle fracture. Table 12.16.1 summarizes these modes.

12.17 Archimedes Principle

Archimedes (250 BC) stated that any object, wholly or partially immersed in a fluid, is buoyed up by a force equal to the weight of the fluid displaced by the object. For ships, this means that the weight of the ship is equal to the water that is displaced. It is why large heavy steel ships can float.

Table 12.16.1 Material failure modes.

Type failure	Limiting material value
Fatigue	Fatigue limit of material
Shear	Ultimate shear strength
Yield	Yield strength
Tensile	Tensile strength
Brittle fracture	Toughness

12.18 Centrifugal Force

The centrifugal force equation is much like the force developed on a string that is swinging a rock in a circle. The force equation simplified is given as:

$$F_{centrifugal} = 28.4^*W^*R^*(\text{rpm}/1{,}000)^2$$

In this equation, W is the weight being swung lb_f, R is the swing radius inches, and rpm is the revolutions per minute.

Many were involved in the development of the equation, but Newton has usually been credited with utilizing this force equation. It is based on the change in acceleration at every point on the circle as (W) is rotated.

12.19 What Is Enthalpy?

Enthalpy (h Btu/lb) is an important concept when dealing with water and steam. Most of the time engineers just use the results of steam tables without really understanding what it is. It's the energy in a pound of a substance such as water or steam.

In equation form $h = U + p^*v$, where $U = m^*c^*T$ and is the internal heat of the substance or system and p and v are pressure and volume and from the ideal gas laws, $p^*v = m^*R^*T$ where R is a gas constant. So enthalpy is a function of temperature just like internal energy. The potential energy is therefore $PE = U = h - p^*v$.

13

A Little About Some Famous Scientists Mentioned in This Book

Some Of My Greatest Contributions Were Due To The Efforts Of Others

In all of the books I've written, I always felt it necessary to comment on the works of the many geniuses whose work was used in the books. It's important to remember that we never do anything in a vacuum. Others have always contributed greatly to the work we do. While we may manipulate basic equations into a form uniquely suited to the problem at hand, we are still using other's work. As has been said, "Some of my greatest contributions were due to the efforts of others." Those persons may have been great scientists or technicians who showed you a better way to do something. So for all those that have helped me I am truly grateful.

After reviewing the translated great works of these scientists I realized that I probably couldn't have used them in that state to solve the engineering problems I had worked. It took other talented people to put them in a form useable by engineers.

13.1   Isaac Newton (1642–1726 AD)

I start with Newton because I am in awe of his abilities and his work. He has contributed greatly to just about every area of engineering. I have worked in as you have seen in some of the examples.

He was a physicist and mathematician and in his day he was called a natural philosopher. I wanted to understand his thinking so I obtained "The Principia: Mathematical Principles of Natural Philosophy" that was some of his major works translated from Latin. Newton and Leibniz both shared credit for individually developing calculus. This was not a friendly sharing. I was surprised to see that in his book, Newton used very little calculus and presented most of his work using geometry in various forms. Some say it was because calculus was new and he didn't want others to find errors in his work. The book would have been much clearer to me if he had used calculus.

Unique Methods for Analyzing Failures and Catastrophic Events: A Practical Guide for Engineers, First Edition. Anthony Sofronas.

Newton formulated the laws of motion and universal gravitation. His laws of motion are three physical laws that were the foundation for classical mechanics. They describe the forces on a body and the resulting motion. He did watch apple's fall and wondered why they fell toward the center of the earth and not sidewise. He didn't come up with the gravitational theory instantaneously and it took him 20 years to develop it. It's a myth that an apple hits him on the head.

Newton was active in many other areas such as optics and mathematical equation development. His law of cooling indicates his activity in the area of heat transfer.

There is much debate on "who is the greatest scientist of all times?" but he is usually cited as one of the top two.

13.2   Daniel Bernoulli (1700–1782 AD)

He was a mathematician and physicist and one of many famous Bernoulli mathematicians. You must mention the first name also when you are talking about the Bernoulli's because there were several generations of prominent ones.

The work I used most is Daniel Bernoulli's work in hydrodynamics. Using the conservation of energy, he showed that as the velocity of a fluid increases, its pressure decreases. This is known as a form of Bernoulli's principle and is essential for understanding flight. He also performed work on the elastic analysis of beams and in the area of statistics and kinetic energies of gases.

13.3   Archimedes of Syracuse (287–212 BC)

Considered to be the greatest mathematician of antiquity, he anticipated calculus by using infinitesimal geometry to develop a range of geometrical theorems. This included the area of a circle, the surface area and volume of a sphere, and the area under a parabola. He designed screw pumps, compound pulleys, and other machines.

Archimedes' principle states that a body immersed in a fluid experiences a buoyant force equal to the weight of the fluid it displaces. This is why large seagoing ships of steel can float.

Archimedes' screw is like the screw in a meat grinder and was used to raise water from a lower level to a higher level and was one of his most important inventions. It is still used today for transporting water and solid because of its simplicity and effectiveness. Processing extruders use an Archimedes' screw.

Archimedes is also said to have designed polished copper parabolic-type mirrors to focus sunlight on ships at sea. This was a weapon said to have been used during the siege of Syracuse and is said to have caused ships to burst into flames.

Later tests showed that this would have been possible since ships of those days were sealed with flammable tar. I would have thought a Captain would have just sailed away from the focused beam. Unfortunately, they still lost the battle and Archimedes was mistakenly killed by a Roman soldier.

13.4 William Rankine (1820–1872 AD)

He was a mechanical engineer and is best known in engineering for his work in thermodynamics and development of the Rankine cycle by which heat engines such as steam engines work. However, he was also the first who used the term "Conservation of Energy," although its principles had been used for years. He proposed the Rankine scale of temperature, an absolute or thermodynamic scale, the degree of which is equal to a Fahrenheit degree. He was one of the founders of the science of thermodynamics.

He performed great works in mechanical engineering as is evident in his book, Manual of Applied Mechanics, 1858, 640 pages. He was one of the first to recognize fatigue failures as cyclic stress problems and he analyzed many actual failures.

13.5 Leonardo da Vinci (1452–1519 AD)

Leonardo was considered an individual with an unquenchable curiosity. This resulted in his expertise in painting, sculpting, anatomy, astronomy, mathematics, and engineering. Historians regarded him as a true Renaissance Man or Universal Genius. One only has to travel to Rome and elsewhere in the world to experience his art and sculptures.

As an engineer, I enjoyed reading about his machines. He was credited with the invention of the parachute, helicopter, and tank. It is said that he was once swinging a ruler over his head for some reason and noticed how his arm rose a little. This had him think of the rotor-type lift motion used in a helicopter. Leonardo did a lot of reading and some of his inventions were based on the work of others. For example, Archimedes screw design and his "fire mirrors" were used in his studies of optics and design of a telescope and welding and soldering equipment.

13.6 Heinrich Holzer

He developed the Holzer method that is a tabular method for calculating both torsional and linear natural frequencies and mode shapes of multi-mass systems. Though more modern approaches are now used, it is still useful to show how a torsional system is analyzed and how the remainder torque must be near zero. Since it is used in this book, its method is outlined in Section 7.6.

13.7   Stephan Timoshenko (1878–1972)

Dr. Stephan Timoshenko, who immigrated from Russia to the United States, commented on the lack of mathematical skills of engineers in the vibration community [1]. He worked for the research division of Westinghouse in the United States for several years setting up training programs and solving practical engineering problems before going into university teaching. In Russia, he was writing articles on the torsional vibration of shafts as early as 1905. He wrote his first book on vibration in 1928 and the third edition is the one I use today.

Reference

1 Timoshenko, S.P. (1968). *As I Remember*. D. Van Nostrand Company, Inc. Translated from Russian by R. Addis.

13.8   Jacob P. Den Hartog (1901–1989)

Dr. Jacob P. Den Hartog was the author of my favorite vibration book [1] and the first edition was written in 1934. Den Hartog mentored under Timoshenko and learned applied vibration from him as he was degreed as an electrical engineer. He developed his knowledge further by solving many practical problems during World War II when new ships, aircraft, and military equipment were being constructed [2].

I once had the privilege of attending a lecture on vibrations at M.I.T. given by Dr. Den Hartog and was amazed at the ease with which he presented difficult subjects. He was in his seventies at that time and I can only imagine his energy when he was in his prime. He was a most entertaining, animated, and enjoyable speaker to listen to.

References

1 Hartog, D. and Pieter, J. *Mechanical Vibration*, 4e. Van Nostrand.
2 Crandall, S.H. (1995). *A Biographical Memoir of Jacob Pieter Den Hartog*. Washington, DC: National Academies Press.

13.9   Wilson, Ker, William

My practical introduction to torsional vibration was through the use of Dr. Ker Wilson's series of books Practical Solution of Torsional Vibration Problems. While the theory of the subject was learned in college, the application on actual ship systems was implemented utilizing my mentors and this series of books. These were used throughout my career and are why they are mentioned here. Especially useful was the calculation of torsional vibration amplitudes and the introduction of the Magnifier Method. This allowed testing data to be applied to future problems.

Dr. Ker Wilson was a prolific writer on torsional vibration theory and problems and his books have been reprinted and referenced many times.

Index

Unique Methods for Analyzing Failures and Catastrophic Events: A Practical Guide for Engineers,
First Edition. Anthony Sofronas.
© 2022 John Wiley & Sons, Inc. Published 2022 by John Wiley & Sons, Inc.